T0305862

Observability and Mathematics

Advances in Applied Mathematics
Series Editors:
Daniel Zwillinger, H. T. Banks

Advanced Engineering Mathematics with MATLAB, 4th Edition
Dean G. Duffy

Quadratic Programming with Computer Programs
Michael J. Best

Introduction to Radar Analysis
Bassem R. Mahafza

CRC Standard Mathematical Tables and Formulas, 33rd Edition
Edited by Daniel Zwillinger

The Second-Order Adjoint Sensitivity Analysis Methodology
Dan Gabriel Cacuci

Operations Research
A Practical Introduction, 2nd Edition
Michael Carter, Camille C. Price, Ghaith Rabadi

Handbook of Mellin Transforms
Yu. A. Brychkov, O. I. Marichev, N. V. Savischenko

Advanced Mathematical Modeling with Technology
William P. Fox, Robert E. Burks

Introduction to Quantum Control and Dynamics
Domenico D'Alessandro

Handbook of Radar Signal Analysis
Bassem R. Mahafza, Scott C. Winton, Atef Z. Elsherbeni

Separation of Variables and Exact Solutions to Nonlinear PDEs
Andrei D. Polyanin, Alexei I. Zhurov

Boundary Value Problems on Time Scales, Volume I
Svetlin Georgiev, Khaled Zennir

Boundary Value Problems on Time Scales, Volume II
Svetlin Georgiev, Khaled Zennir

Observability and Mathematics
Fluid Mechanics, Solutions of Navier-Stokes Equations, and Modeling
Boris Khots

Handbook of Differential Equations, 4th Edition
Daniel Zwillinger, Vladimir Dobrushkin

Experimental Statistics and Data Analysis for Mechanical and Aerospace Engineers
James Middleton

https://www.routledge.com/Advances-in-Applied-Mathematics/book-series/CRCADVAPP
MTH?pd=published,forthcoming&pg=1&pp=12&so=pub&view=list

Observability and Mathematics

Fluid Mechanics, Solutions of Navier-Stokes Equations, and Modeling

Boris Khots

CRC Press
Taylor & Francis Group
Boca Raton London New York

CRC Press is an imprint of the
Taylor & Francis Group, an **informa** business

A CHAPMAN & HALL BOOK

First edition published 2021
by CRC Press
6000 Broken Sound Parkway NW, Suite 300, Boca Raton, FL 33487-2742

and by CRC Press
2 Park Square, Milton Park, Abingdon, Oxon, OX14 4RN

© 2022 Boris Khots

CRC Press is an imprint of Taylor & Francis Group, LLC

Reasonable efforts have been made to publish reliable data and information, but the author and publisher cannot assume responsibility for the validity of all materials or the consequences of their use. The authors and publishers have attempted to trace the copyright holders of all material reproduced in this publication and apologize to copyright holders if permission to publish in this form has not been obtained. If any copyright material has not been acknowledged please write and let us know so we may rectify in any future reprint.

Except as permitted under U.S. Copyright Law, no part of this book may be reprinted, reproduced, transmitted, or utilized in any form by any electronic, mechanical, or other means, now known or hereafter invented, including photocopying, microfilming, and recording, or in any information storage or retrieval system, without written permission from the publishers.

For permission to photocopy or use material electronically from this work, access www.copyright.com or contact the Copyright Clearance Center, Inc. (CCC), 222 Rosewood Drive, Danvers, MA 01923, 978-750-8400. For works that are not available on CCC please contact mpkbookspermissions@tandf.co.uk

Trademark notice: Product or corporate names may be trademarks or registered trademarks and are used only for identification and explanation without intent to infringe.

ISBN: 978-1-032-00813-4 (hbk)
ISBN: 978-1-032-11856-7 (pbk)
ISBN: 978-1-003-17590-2 (ebk)

DOI: 10.1201/9781003175902

Typeset in CMR10
by KnowledgeWorks Global Ltd.

Dedicated to my children
Natalia, Maria, Alexandr, Dmitriy

Contents

Foreword

The existence and smoothness of Navier-Stokes equations is one of the most challenging problems in contemporary mathematics and physics, specifically in three-dimensional space. Theoretical physicists and mathematicians are involved in research of Navier-Stokes equations due its close relationship with the Euler equation. However, understanding of solutions of equations describing fluid motion motivates not just theoretical physicists, but practitioners as well – especially those in maritime and air transport business - predicting location of rogue waves or turbulence are critical to saving lives. What remains highly surprising is the fact that while Navier-Stokes and Euler equations were created quite a few years ago, our understanding of them remains minimal.

This book provides a glimpse of hope into solving these challenges by reinventing the fundamental mathematical logic used to derive most of these concepts. The book essentially remodels classic fluid mechanics concepts and derives equation of continuity, Euler equation of fluid motion, energy flux and momentum flux equations, Navier-Stokes equations from Mathematics with Observers point of view, and further more provides an approach to solution of classic problems with this interpretation of fluid motion laws.

Mathematics with Observers in a nutshell is a reimagining of the very basic fundamentals of arithmetic, geometry, and calculus that rejects the notion of infinity and recalibrates all physical calculations into a nested set of observable arithmetic that locally coincides with traditional mathematics, but has highly specific nuances that change the dynamics of physics research. This book provides a fresh perspective into classical problems and offers an alternative means of understanding fluid mechanics. It does require the reader to understand Mathematics with Observers concept, so it is highly recommended to review previously published work in order to enable ease of understanding of the material.

Dmitriy Khots, iMath Consulting LLC, Omaha, Nebraska

Preface

This book describes author's approach to an old classic problem – the existence of solutions of Navier-Stokes equations. The existence and smoothness of such solutions is one of the most challenging and hard problems in contemporary mathematics and physics. The main objective of the book is to model and derive of equation of continuity, Euler equation of fluid motion, energy flux and momentum flux equations, Navier-Stokes equations from Mathematics with Observers point of view, and approach to solution classic problem for this interpretation of fluid motion laws.

Mathematics with Observers were developed by authors based on denial of infinity idea. Let's note the basic mathematical parts such that arithmetic, linear algebra, mathematical analysis (calculus), geometry, differential geometry, algebra, Lie groups and Lie algebras, and functional analysis are based on infinity idea. That's about such concepts as limit, derivative, integral, continuity, line, plane, space, manifold, and many others. Discussion of the infinity idea is not new in mathematics and physics, and takes a place in works of David Hilbert, Peter Rashevsky, Kurt Godel, and Gerhard Gentzen. There are few works of Esenin-Volpin who was a supporter of ultrafinite point of view, and the direction in Mathematical Logic with name intuitionism of Brouwer's school, where one considers as legitimate only finite sets. And in our time discussion of an infinity idea is still continuing. That's about the nine axioms, collectively called "ZFC" (a nine-item list of rules called Zermelo-Fraenkel set theory with the axiom of choice), and H. Woodin "V=ultimate L" axiom. There is well-known set-theoretical and topological result – straight line and any as much as desired small open interval of this line are homeomorphic, i.e. have the identical topological structure. And it is clear that this result contradicts to atomic structure of nature. Many of these analogous situations pushed authors for Mathematics with Observers development. So, Mathematics with Observers were developed by authors based on denial of infinity idea. We introduce Observers into arithmetic, and arithmetic becomes dependent on Observers. And after that the basic mathematical parts also become dependent on Observers.

Considering Mathematics with Observers applications to Fluid Mechanics, we first explain the concept of Mathematics with Observers and it appears to be trivial, but it is highly nuanced and that is what makes all the difference. In addition, we consider classic fluid mechanics equations as a first approach to nature. But then, when we consider the physical conservation laws of mass, momentum, and energy from Mathematics with Observers point of view, that is what allows us to see much richer details in the nature of Fluid Mechanics.

The author would like to thank Bob Ross for invitation to write this book and for many valuable advices including the title of this book. Also, the author wishes to express thanks to Dmitriy Khots for his help in each step and detail of this book, Ilya Markevich for helping with some LaTex complications. The author thanks Lev Yakubovich for his help with derivative sample calculations. The author would also like to thank Vaishali Singh, Mansi Kabra, and Ashraf Reza for their help in editing and publishing this book.

Contributors

Dmitriy Khots
iMath Consulting LLC
Omaha, Nebraska

1

Introduction

The Mathematics with Observers was developed by authors based on denial of infinity idea. Let's note the basic mathematical parts such that arithmetic, linear algebra, mathematical analysis (calculus), geometry, differential geometry, algebra, Lie groups and Lie algebras, functional analysis are based on infinity idea. That's about such concepts as limit, derivative, integral, continuity, line, plane, space, manifold and many others. Discussion of the infinity idea is not new in mathematics and physics. We see this discussion in works of many mathematicians and physicists, for example David Hilbert, Peter Rashevsky, and the direction in Mathematical Logic with name intuitionism of Brouwer's school, where one considers as legitimate only finite sets. We know the infinite appears in the infinite number sequences that define the real numbers. In traditional science, we always assume continuity of matter, e.g., a block of wood or a pool of water can be divided into pieces infinitely, where each resulting chunk has the same divisibility properties, however, every experiment performed in physics always has a limit to divisibility due to the nature of measurement mechanisms deployed in the real world. There is well known set-theoretical and topological result – straight line and any as much as desired small open interval of this line are homeomorphic. And it is clear that this result contradicts to atomic structure of nature. Many of these analogous situations pushed authors for Mathematics with Observers development.

We introduce Observers into arithmetic, and arithmetic becomes dependent on Observers. We consider Observer dependent ascending chain of embedded sets of decimal fractions and their Cartesian products. For every set, arithmetic operations are defined (these operations locally coincide with standard operations until the results of these operations are touching of corresponding sets borders). The system of observers becomes a finite well-ordered system. The observers are ordered by their level of "depth", i.e. an observer with greater depth sees any observer with smaller depth, and inverse an observer with a given depth is unaware of observers with larger depth values. Probability and stochastic appear automatically because of these sets, observers and arithmetic. In difference of classic arithmetic in Mathematics with Observers additive and multiplicative associativities, distributivity may fail; multiplicative inverses do not necessarily exist, or if they do, they are not necessarily unique; square roots do not necessarily exist; there are zero divisors. And the probabilities of additive and multiplicative associativities, distributivity, multiplicative inverses existence and uniqueness, square roots existence are less

DOI: 10.1201/9781003175902-1

than 1. And after that the basic mathematical parts also become dependent on Observers.

In this book we consider Mathematics with Observers applications to Fluid Mechanics. We follow classic Physics interpretation of Fluid Mechanics laws and contemporary problems. Based on Physics data we developed here Mathematics with Observers interpretation of the main laws of Fluid Mechanics. This new equations are derived from the basic principles of continuity of mass, momentum, and energy. We get here new thermodynamical equations, continuity equation, Euler equation of motion of the fluid, energy flux and moment flux equations, incompressible fluids equations, Navier-Stokes equations. All these equations become stochastic because depend on observers point of view.

We consider thermal natural variables: temperature and entropy, and mechanical natural variables: pressure and volume. Also we consider thermodynamic potentials as functions of their thermal and mechanical variables: the internal energy with given temperature and pressure considered as the parameters, enthalpy with given temperature and volume considered as the parameters, Helmholtz free energy with given entropy and pressure considered as the parameters, and Gibbs free energy with given entropy and volume considered as the parameters. We consider classic thermodynamical equations from Mathematics with Observers point of view. And we prove that they are correct with probabilities less than 1 in Mathematics with Observers because applied here the main Observer's tools (in particular specific properties of differentiation and integration) give these results. And thermodynamical equations in Mathematics with Observers become stochastic.

We consider a finite arbitrary volume, called a control volume, over which the basic principles of continuity of mass, momentum, and energy can be applied. The control volume can remain fixed in space or can move with the fluid. We take the ideal fluid velocity and two thermodynamic quantities - pressure and density. We consider the volume and mass of fluid in this volume and calculate the total mass of fluid flowing in and out of this volume in unit time. We calculate decrease per unit time in the mass of fluid in this volume and get the equation of continuity. Using this classic way and applying here Mathematics with Observers tools we are remodeling fluid continuity and get the equation of continuity in Mathematics with Observers. And equation of continuity in Mathematics with Observers becomes stochastic. We proved that classic equation of continuity takes a place with probability less than 1.

For modeling the motion of the ideal fluid we take pressure as a function defined on cube of the ideal fluid and calculate the total vector-force acting on this volume. By Newton second law we get the equation of motion of the volume element in the fluid. Using this classic way and applying here Mathematics with Observers tools we are remodeling motion of the ideal fluid and get this equation of the motion in Mathematics with Observers. So we get Euler equation of motion of the fluid in Mathematics with Observers. If the fluid is in gravitational field, we add an additional force and get Euler equation in Mathematics with Observers with this additional force. We get additional view of the Observer's Euler equation of motion of the fluid, in which it involves only the velocity. Also we rewrite Observer's Euler equation of motion of the fluid in gravitational field. And we write the Observer's Euler equation for the fluid at rest in uniform gravitational field. And Euler equation of motion of the fluid in Mathematics with Observers becomes stochastic. We proved that classic Euler equation takes a place with probability less than 1.

We take some volume element fixed in space and define the energy of unit volume of ideal fluid as a sum of kinetic energy and the internal energy. We calculate also the change in this energy. And finally we get energy flux equation. Using this classic way and applying here Mathematics with Observers tools we are remodeling energy picture and get Mathematics with Observers energy flux equation. We calculate the momentum of unit volume of ideal fluid and the rate of this momentum change. And finally we get momentum flux equation. In process of calculations we consider the momentum flux density matrix. Using this classic way and applying here Mathematics with Observers tools we are remodeling momentum picture and get Mathematics with Observers momentum flux equation. In particular in classic Fluid Mechanics the momentum flux density matrix is called the momentum flux density tensor, but in Mathematics with Observers it is not a tensor in classic linear algebra sense, but only matrix in the chosen coordinate system. And energy flux and moment flux equations in Mathematics with Observers becomes stochastic. We proved that classic energy flux and moment flux equations take a place with probability less than 1.

If the density of flow of liquids or gases is invariable, i.e constant throughout the volume of the fluid and throughout its motion – it is named the incompressible flow. Euler equation in Mathematics with Observers does not change in this case. But equation of continuity in Mathematics with Observers becomes more simple. And we get from equation of continuity in Mathematics with Observers the equation with Laplace's operator of potential. This equation must be supplemented by boundary conditions at the surfaces where the fluid meets solid bodies. We consider also the situation of plane flow, i.e. the velocity distribution in a moving fluid depends on only two coordinates. We introduce the complex potential in Mathematics with Observers and the complex velocity and find corresponding equation. And all incompressible fluids equations in Mathematics with Observers becomes stochastic. We proved that classic incompressible fluids equations take a place with probability less than 1.

Instead of ideal fluids we consider viscous fluids. The viscosity (internal friction) causes another, irreversible, transfer of momentum from points where the velocity is large to those where it is small. The equation of motion of a viscous fluid may therefore be obtained by adding to "ideal" momentum flux some term which gives the irreversible "viscous" transfer of momentum in the fluid. The equation of motion of a viscous fluid can now be obtained by simply adding some expression to the right hand side of Euler equation in Mathematics with Observers. And we get Navier-Stokes equations in Mathematics with

Observers. We get these equations in various situations with fluid and any given externally applied force. And all these equations in Mathematics with Observers becomes stochastic. We proved that classic Navier-Stokes equations take a place with probability less than 1. We make analysis of Navier-Stokes equations in Mathematics with Observers. And we make analysis of the solution existence of these equations.

Also we consider here the interpretation of Relativistic Fluid Mechanics in Mathematics with Observers. The governing principles in Fluid Mechanics are the conservation laws for mass, momentum, and energy. And in classic Physics and Mathematics the conservation laws characterizing special relativistic fluid mechanics are invariant (in fact co-variant) under Poincare group of transformations. We consider this situation from Mathematics with Observers point of view. First of all we prove that the sets of all invertible 3×3, 4×4, 5×5 matrices are not the Lie groups in Mathematics with Observers, and group definition's conditions take a place here with some probability less than 1. Also we prove that the sets of all orthogonal matrices $\mathbf{O(3)}$ is not the Lie groups in Mathematics with Observers, and group definition's conditions take a place here with some probability less than 1. And we prove that the sets of all Lorentz matrices $\mathbf{L}$ is not the Lie groups in Mathematics with Observers, and group definition's conditions take a place here with some probability less than 1. And finally we prove that the sets of all Poincare matrices $\mathbf{P}$ is not the Lie groups in Mathematics with Observers, and group definition's conditions take a place here with some probability less than 1. That means in Mathematics with Observers the probabilities of the conservation laws characterizing special relativistic fluid mechanics are invariant (in fact co-variant) under Poincare group of transformations are less than 1.

2

Observability and Arithmetic

We follow now the ideas of David Hilbert (see [3]), Peter Rashevsky (see [9]) and describe the Mathematics with Observers (see [5], [4]) based on denial of infinity idea.

Let's consider the set W_n of all decimal fractions, such that there are at most n digits in the integer part and n digits in the decimal part of the fraction. Visually, an element in W_n looks like

$$\pm \underbrace{\_ \ \cdots \ \_}_{n} . \underbrace{\_ \ \cdots \ \_}_{n} .$$

We have $W_n \subseteq W_k$, if $n \le k$, $k, n = 2, 3, \ldots$.

We call the "W_n-observer" somebody (or some system) who *deals* with W_n.

The system of observers becomes a finite well-ordered system. The observers are ordered by their level of "depth", i.e. each "W_n-observer" has a depth number n, and an observer with depth k sees any observer with depth $n \le k$.

Moreover, an observer with a given depth is unaware of observers with larger depth values.

So, the "W_k-observer" is the abbreviation for somebody (some system) who *deals* with W_k, sees observers with depth $n \le k$ and does not see observers with depth $n > k$.

When we talk about observers, we shall always have some fixed observer (called 'us') with the possible greatest depth value who oversees all others.

Let $n = 2$, so we are in W_2. Here are some examples of elements of W_2: $3.14, -99, 0.1 \in W_2$ and $0.115, 123.9, -100000 \notin W_2$.

Let's consider now the set W_n of all elements $a = a_0.a_1 \ldots a_n$ where the integer part can be written as $a_0 = b_{n-1} \ldots b_0$, where $b_{n-1}, \ldots, b_0, a_1, \ldots, a_n \in \{0, 1, \ldots, 9\}$.

Now, given $c = \pm c_0.c_1 \ldots c_n$, $d = \pm d_0.d_1 \ldots d_n \in W_n$ we endow W_n with the following arithmetic $(+_n, -_n, \times_n, \div_n)$ – from W_m – observer point of view $(m \ge n)$.

Let's define first addition and subtraction in W_n:

$$c \pm_n d = \begin{cases} c \pm d, & \text{if } c \pm d \in W_n \\ \text{not defined}, & \text{if } c \pm d \notin W_n \end{cases}$$

where $c, d \in W_n$, $c \pm d$ is the standard addition and subtraction (in classic arithmetic).

DOI: 10.1201/9781003175902-2

Now, the examples of addition and subtraction in W_2:

$$2.08 +_2 11.9 = 13.98$$

$$(-2.08) +_2 11.9 = 9.82$$

$$80 +_2 24 = \text{not defined}$$

$$21.36 -_2 0.87 = 20.49$$

$$1.36 -_2 16.95 = -15.59$$

$$1.36 -_2 (-99.95) = \text{not defined}$$

Let's define now multiplication in W_n.

For given non-negative $c = c_0.c_1c_2\ldots c_{n-1}c_n$, $d = d_0.d_1d_2\ldots d_{n-1}d_n \in W_n$ we put

$c \times_n d = (((\ldots((c_0 \cdot d_0.d_1d_2\ldots d_{n-1}d_n) + 0.c_1 \cdot d_0.d_1d_2\ldots d_{n-1}) + \ldots) + {}$
$+ 0.0\ldots 0c_{n-1} \cdot d_0.d_1) + 0.0\ldots 0c_n \cdot d_0)$

or we can write down

$$c \times_n d = \sum_{k=0}^{n} {}^n \sum_{m=0}^{n-k} {}^n 0.\underbrace{0\ldots 0}_{k-1} c_k \cdot 0.\underbrace{0\ldots 0}_{m-1} d_m$$

where $c, d \geq 0$, $c_0 \cdot d_0 \in W_n$, $0.\underbrace{0\ldots 0}_{k-1} c_k \cdot 0.\underbrace{0\ldots 0}_{m-1} d_m$ is the standard multiplication (in classic arithmetic), and $k = m = 0$ means that $0.\underbrace{0\ldots 0}_{k-1} c_k = c_0$ and

$0.\underbrace{0\ldots 0}_{m-1} d_m = d_0$.

If either $c < 0$ or $d < 0$, then we compute $|c| \times_n |d|$ and define $c \times_n d = \pm |c| \times_n |d|$, where the sign $\pm$ is defined as usual (in classic arithmetic). Note, if the content of at least one parenthesis (in previous formula) is not in W_n, then $c \times_n d$ is not defined. And we write

$$((\ldots (f_1 +_n f_2)\ldots) +_n f_N) = \sum_{i=1}^{N} {}^n f_i$$

for $f_1, \ldots, f_N$ iff the contents of any parenthesis are in W_n, $f_1, \ldots, f_N \in W_n$.

Now, the examples of multiplication in W_2:

$$11 \times_2 8 = 88$$

$$(-5) \times_2 19 = -95$$

$$11 \times_2 12 = \text{not defined}$$

$$3.41 \times_2 2.64 = 8.98$$

$$3.41 \times_2 (-2.64) = -8.98$$

$$3.41 \times_2 42.64 = \text{not defined}$$

$$99.41 \times_2 1.64 = \text{not defined}$$

$$0.85 \times_2 0.02 = 0$$

And finally let's define division in W_n.

For given $c, d \in W_n, d \neq 0$

$c \div_n d = r$, if $d \times_n r = c$.

If such r does not exist, $c \div_n d$ is not defined

Now, the examples of division in W_2:

$$80 \div_2 4 = 20$$

$$1 \div_2 0.5 = \{2, 2.01, 2.02, 2.03, 2.04, 2.05, 2.06, 2.07, 2.08, 2.09\}$$

So, we get 10 different r's.

$$1 \div_n 3 = \text{not defined}$$

since no r exists because

$$3 \times_2 0.33 = 0.99; 3 \times_2 0.34 = 1.02$$

We will use many times in this book the notion "probability".

We use notion "probability" in the following sense. We consider the set of events in W_n or in Cartesian products $W_n \times W_n$ or $W_n \times W_n \times \ldots \times W_n$ from W_m- observer point of view ($m \geq n$) assuming that W_m- observer may see full set of events. For example, the set of all points in W_2 may see W_m- observer with $m \geq 5$. The set of all points in $W_2 \times W_2$ may see W_m- observer with $m \geq 9$. The set of all points in $W_2 \times W_2 \times W_2$ may see W_m- observer with $m \geq 13$. Let's number of elements of the full set of events is S from W_m- observer point of view. And let's number of elements of the specific set of events, probability of which we are defining, is R from W_m- observer point of view. We call the "probability" of these specific events from W_m- observer point of view the number

$$p = p_{nm} \in W_m$$

such that

$$|S \times_m p -_m R| = min_p$$

Here we provide some basic examples to illustrate what might happen with standard properties of classic arithmetic.

1. Additive associativity may fail:

$$(x +_n y) +_n z \neq x +_n (y +_n z)$$

E.g. let $10, 95, -35 \in W_2$, then $10 +_2 95 \notin W_2$, hence

$$(10 +_2 95) +_2 (-35) \notin W_2$$

and

$$10 +_2 (95 -_2 35) = 70 \in W_2$$

But for $1, 2, 3 \in W_2$ we have

$$1 +_2 (2 +_2 3) = (1 +_2 2) +_2 3 = 6 \in W_2$$

So, we get a theorem:

Theorem 2.1 *The probability of equality*

$$(x +_n y) +_n z = x +_n (y +_n z)$$

is less than 1.

2. Multiplicative associativity may fail:

$$(x \times_n y) \times_n z \neq x \times_n (y \times_n z)$$

E.g. let $50.12, 0.85, 0.61 \in W_2$, then

$$50.12 \times_2 0.85 = 42.58; (50.12 \times_2 0.85) \times_2 0.61 = 25.92$$

whereas

$$0.85 \times_2 0.61 = 0.48; 50.12 \times_2 0.48 = 24.04$$

But for $1, 2, 3 \in W_2$ we have

$$1 \times_2 2 = 2; (1 \times_2 2) \times_2 3 = 6.00$$

$$2 \times_2 3 = 6; 1 \times_2 (2 \times_2 3) = 6.00$$

So, we get a theorem:

Theorem 2.2 *The probability of equality*

$$(x \times_n y) \times_n z = x \times_n (y \times_n z)$$

is less than 1.

3. Distributivity may fail:

$$x \times_n (y +_n z) \neq x \times_n y +_n x \times_n z$$

E.g. let $1.81, 0.74, 0.53 \in W_2$, then

$$0.74 +_2 0.53 = 1.27; 1.81 \times_2 1.27 = 2.24; 1.81 \times_2 0.74$$
$$= 1.3; 1.81 \times_2 0.53 = 0.93$$

so that

$$1.81 \times_2 0.74 +_2 1.81 \times_2 0.53 = 2.23 \neq 2.24$$

But for $1, 2, 3 \in W_2$ we have

$$1 \times_2 (2 +_2 3) = 1 \times_2 2 +_2 1 \times_2 3 = 5.00$$

So, we get a theorem:

Theorem 2.3 *The probability of equality*

$$x \times_n (y +_n z) = x \times_n y +_n x \times_n z$$

is less than 1.

4. Lack of the distribution law leads to the following result: The statement "$x|y$ and $x|z \Rightarrow x|(y+z)$" is false. Here $a|b \Leftrightarrow \exists r : a \times_n r = b$. Assume that $x|y; x|z; x, y, z \in W_2$. Let

$$x = 0.17; r_1 = 0.85; r_2 = 0.63$$

$$y = 0.17 \times_2 0.85 = 0.08$$

$$z = 0.17 \times_2 0.63 = 0.06$$

Then $y +_2 z = 0.14$. And

$$0.17 \times_2 0.99 = 0.09 < 0.14$$

$$0.17 \times_2 1 = 0.17 > 0.14$$

But for $x = 2, y = 4, z = 6 \in W_2$ we have

$$y = 2 \times_2 x; z = 3 \times_2 x; y +_2 z = 5 \times_2 x$$

So, we get a theorem:

Theorem 2.4 *The probability of statement "$x|y$ and $x|z \Rightarrow x|(y+z)$" correctness is less than 1.*

5. Multiplicative inverses do not necessarily exist, or if they do, they are not necessarily unique in W_n. Here are some examples: let $2 \in W_n$, then $0.5 \in W_n$ is the unique inverse of 2 for any $W_n, n \geq 2$. On the other hand, 3 will not have an inverse in any W_n. Now, let $2^{-1} = 0.5$, then $(0.5)^{-1}$ is actually the following set

$$\{2, 2.01, 2.02, 2.03, 2.04, 2.05, 2.06, 2.07, 2.08, 2.09\} \in W_2$$

Therefore, $(2^{-1})^{-1}$ is not necessarily 2, hence all we can claim is that if x^{-1} exists, then

$$x \in \left\{ \left(x^{-1}\right)^{-1} \right\}$$

Further, if an inverse of an element exists in W_n, it does not necessarily exist in W_m for $m \neq n$, independent of the order of m and n, e.g. if $0.91 \in W_2$, then

$$(0.91)^{-1} = \{1.1, 1.11, 1.12, 1.13, 1.14, 1.15, 1.16, 1.17, 1.18, 1.19\} \in W_2$$

but $(0.91)^{-1} \notin W_4$, on the other hand, $16^{-1} = 0.0625 \in W_4$, but $16^{-1} \notin W_2$.

6. Square roots do not necessarily exist. Some examples are, if $4 \in W_n$, then $\sqrt{4} = 2$ for any n and $\sqrt{3}$ does not exist in $n = 2$, because

$$1.75 \times_2 1.75 = 2.99; 1.76 \times_2 1.76 = 3.01$$

Further, if a square root of an element exists in W_n, it does not necessarily exist in W_m for $m \neq n$, independent of the order of m and n, e.g. $\sqrt{2} = 1.42 \in W_2$, since $1.42 \times_2 1.42 = 2$, but $\sqrt{2} \notin W_4$, since

$$1.4143 \times_4 1.4143 = 1.9999; 1.4144 \times_4 1.4144 = 2.0001$$

Also, $\sqrt{1.01} = 1.005 \in W_4$, but $\sqrt{1.01} \notin W_2$, since

$$1 \times_2 1 - 1; 1.01 \times_2 1.01 = 1.02$$

Next, the following basic theorem can be stated for W_n.

Theorem 2.5 *Any W_n has zero divisors. For example:*

$$0.\underbrace{0...0}_{n-1}1 \times_n 0.\underbrace{0...0}_{n-1}1 = 0.$$

For all $x, y \in W_n$ with $x, y \geq 0$, $x - y \in W_n$.
If $x, y, t, u \in W_n$ and $x \geq t \geq 0$ and $y \geq u \geq 0$ and $x \times_n y \in W_n$, then $t \times_n u \in W_n$ and $x \times_n y \geq t \times_n u$.
If given $a \in W_n$ such that there is a unique $a^{-1} \in W_n$, then $|a| \geq 1$.
If $|a| < 1$ and a^{-1} exists, then $\left|\{a^{-1}\}\right| > 1$.
If $\left|\{a^{-1}\}\right| > 1$, then $|a| < 1$.

Let's consider now additional valuable properties of introduced arithmetic. Standard multiplications identities become wrong, for example

$$(x + y)^2 \neq x^2 + 2(xy) + y^2.$$

We have

Theorem 2.6

$$P\left((a +_n b) \times_n (a +_n b)\right) = (a \times_n a +_n 2 \times_n (a \times_n b)) +_n b \times_n b) < 1$$

where P is a probability.

We can see a proof below.
First let's $n = 2$. Then

1. Left side of equality is $(1.32 +_2 2.43) \times_2 (1.32 +_2 2.43) = 3.75 \times_2 3.75 = 13.99$, right side consists from two parts. First, $1.32 \times_2 1.32 = 1.73$; second, $2 \times_2 (1.32 \times_2 2.43) = 6.38$, and finally $2.43 \times_2 2.43 = 5.88$. That means $1.73 +_2 6.38) +_2 5.88 = 13.99$. I.e left side equals to right. But now let's consider the following calculations.

2. Left side of equality is $(1.32 +_2 2.79) \times_2 (1.32 +_2 2.79) = 4.11 \times_2 4.12 = 16.89$, right side consists of two parts. First, $1.32 \times_2 1.32 = 1.73$; second, $2 \times_2 (1.32 \times_2 2.79) = 7.28$, and finally $2.79 \times_2 2.79 = 7.65$. That means $1.73 +_2 7.28 +_2 7.65 = 16.66$. I.e left side does not equal to right.

Let's consider now a random variable

$$\delta_1 = (a +_n b) \times_n (a +_n b) - ((a \times_n a +_n 2 \times_n (a \times_n b)) +_n (b \times_n b))$$

where δ_1 and all elements of right side belong to W_n.

General proof for W_n you can see below.

If a, b non-negative integers in W_n and $(a +_n b) \times_n (a +_n b) \in W_n$, then $\delta_1 = 0$. Let's consider now $a = 0.\underbrace{9 \ldots 9}_{n}$ and $b = 0.\underbrace{0 \ldots 08}_{n}$. Then $a +_n b = 1.\underbrace{0 \ldots 07}_{n}$ and $(a +_n b) \times_n (a +_n b) = 1.\underbrace{0 \ldots 07}_{n} \times_n 1.\underbrace{0 \ldots 07}_{n} = 1.\underbrace{0 \ldots 014}_{n}$, but $a \times_n a < 1$, $b \times_n b = 0$, and $2 \times_n (a \times_n b) = 0$. I.e. $\delta_1 \neq 0$.

We have also the following theorem.

Theorem 2.7

$$P(c \times_n (a +_n b) = c \times_n a +_n c \times_n b) < 1$$

where P is a probability.

Below you can see a proof.

First Let's $n = 2$. Then

1. Left side of equality is $2 \times_2 (3 +_2 6) = 2 \times_2 9 = 18$, right side consists of two parts. First, $2 \times_2 3 = 6$, then $2 \times_2 6 = 12$ and $6 +_2 12 = 18$ I.e. left side equals to right. But go to next calculations.

2. Left side of equality is $2.41 \times_2 (3.14 +_2 0.58) = 2.41 \times_2 3.72 = 8.95$, right side consists of two parts. First, $2.41 \times_2 3.14 = 7.55$, then $2.41 \times_2 0.58 = 1.36$ $7.55 +_2 1.36 = 8.91$. I.e. left side does not equal to right.

Let's consider a random variable

$$\delta_2 = c \times_n (a +_n b) -_n (c \times_n a +_n c \times_n b)$$

where δ_2 and all elements of right side belong to W_n.

General proof for W_n you can see below.

If a, b, c – non-negative integers in W_n and $a \times_n (b \times_n c) \in W_n$, then $\delta_2 = 0$. Let's consider now $a = 2$, $b = 0.\underbrace{9 \ldots 9}_{n}$ and $c = 0.\underbrace{0 \ldots 01}_{n}$. Then $b \times_n c = 0$, $a \times_n (b \times_n c) = 0$, $a \times_n b = 1.\underbrace{9 \ldots 98}_{n}$, and $(a \times_n b) \times_n c = 0.\underbrace{0 \ldots 01}_{n}$. I.e. $\delta_2 \neq 0$.

We have also the following theorem.

Theorem 2.8 *Let's*

$$\delta_3 = c \times_n (a \times_n b) -_n (c \times_n a) \times_n b$$

Then $0 < P(\delta_3 = 0) < 1$, where P is a probability.

You can see a proof of this theorem below.

First Let's $n = 2$. Then

1. Left side of this equality is $2 \times_2 (3 \times_2 6) = 2 \times_2 18 = 36$, right side consists from two parts . First, $2 \times_2 3 = 6$, then $6 \times_2 6 = 36$. I.e left side equals to right. But let's consider the following calculations.

2. Left side is $2.41 \times_2 (3.14 \times_2 0.58) = 2.41 \times_2 1.79 = 4.27$, for right side we get first $2.41 \times_2 3.14 = 7.55$, then $7.55 \times_2 0.58 = 4.31$. And left side does not equal to right.

Let's consider a random variable

$$\delta_3 = c \times_n (a \times_n b) -_n (c \times_n a) \times_n b$$

where δ_3 and all elements of right side belong to W_n.

General proof for W_n you can see below.

If a, b, c are non-negative integers in W_n and

$$c \times_n (a \times_n b) \in W_n$$

then $\delta_3 = 0$. Let's consider $c = 2$, $a = 0.\underbrace{9\ldots99}_{n}$ and $b = 0.\underbrace{0\ldots01}_{n}$. Then

$$\delta_3 = 2 \times_n (0.9\ldots99 \times_n 0.0\ldots01) -_n (2 \times_n 0.9\ldots99) \times_n 0.0\ldots01$$

$$= 0 -_n 0.0\ldots01 = -0.0\ldots01 \neq 0$$

Let's introduce now a complex numbers in Mathematics with Observers. We consider the sets $CW_n, n = 2, 3, \ldots$ – the sets of all formal sums

$$z = a +_n ib$$

where $a, b \in W_n$ and i is formal element, with the following arithmetics:

$$z_1 +_n z_2 = (a_1 +_n ib_1) +_n (a_2 +_n ib_2) = (a_1 +_n a_2) +_n i(b_1 +_n b_2)$$

if $a_1 +_n a_2, b_1 +_n b_2 \in W_n$;

$$z_1 \times_n z_2 = (a_1 +_n ib_1) \times_n (a_2 +_n ib_2) = (a_1 \times_n a_2 -_n b_1 \times_n b_2)$$
$$+_n i(a_1 \times_n b_2 +_n a_2 \times_n b_1)$$

if

$$a_1 \times_n a_2, b_1 \times_n b_2, a_1 \times_n a_2 -_n b_1 \times_n b_2, a_1 \times_n b_2, a_2 \times_n b_1, a_1$$
$$\times_n b_2 +_n a_2 \times_n b_1 \in W_n$$

We have in CW_n:

$$W_n \subset CW_n$$

(we put $a \equiv a +_n i0$),

$$i \times_n i = -1$$

(we put $i \equiv 0 +_n i1$),

$$z_1 +_n z_2 = z_2 +_n z_1$$

$$z_1 \times_n z_2 = z_2 \times_n z_1$$

Based on arithmetic properties in W_n we get

$$(0.01 +_n i0.01) \times_n (0.01 +_n i0.01) = 0$$

(we put $0 \equiv 0 +_n i0$).

So, we have a theorem:

Theorem 2.9 *Any CW_n has zero divisors.*

Also we get the following theorems:

Theorem 2.10

$$P(z_1 +_n (z_2 +_n z_3) = (z_1 +_n z_2) +_n z_3) < 1$$

(P is probability).

Theorem 2.11

$$P(z_1 \times_n (z_2 +_n z_3) = z_1 \times_n z_2 +_n z_1 \times_n z_3) < 1$$

Theorem 2.12

$$P(z_1 \times_n (z_2 \times_n z_3) = (z_1 \times_n z_2) \times_n z_3) < 1$$

For $z = a +_n ib$ we call absolute value of z the following expression $|z| = \sqrt{a^2 +_n b^2}$, if $a^2 = a \times_n a, b^2 = b \times_n b, a^2 +_n b^2 \in W_n$ and $\sqrt{a^2 +_n b^2}$ exists.

Let's $z_1 = z_2 = 1$, then $|z_1| = |z_2| = 1$ and

$$|z_1 \times_n z_2| = |z_1| \times_n |z_2|$$

Let's now $z_1 = 1 +_2 i$, $z_2 = 1 +_2 i2$, then $z_1 \times_2 z_2 = -1 +_2 i3$ and $|z_1| = \sqrt{2} = 1.42$, $|z_2| = \sqrt{5} = 2.24$, $|z_1 \times_2 z_2| = \sqrt{10}$ does not exist.

Theorem 2.13

$$P(|z_1 \times_n z_2| = |z_1| \times_n |z_2|) < 1$$

3

Observability and Vector Algebra

Let's consider Cartesian product of m copies of W_n: $E_m W_n = \underbrace{W_n \times \ldots \times W_n}_{m}$.

We call "vector" any element from $E_m W_n$: $\mathbf{a} = (a_1, \ldots, a_m)$, $a_1, \ldots, a_m \in W_n$.

If $\mathbf{a}, \mathbf{b} \in E_m W_n$, $\mathbf{a} = (a_1, \ldots, a_m)$, $\mathbf{b} = (b_1, \ldots, b_m)$, $\alpha \in W_n$, we define

$$\mathbf{a} +_n \mathbf{b} = (a_1 +_n b_1, \ldots, a_m +_n b_m)$$

if $a_1 +_n b_1, \ldots, a_m +_n b_m \in W_n$.

$$\alpha \times_n \mathbf{a} = (\alpha \times_n a_1, \ldots, \alpha \times_n a_m)$$

if $\alpha \times_n a_1, \ldots, \alpha \times_n a_m \in W_n$.

For $m = 3$ we get standard basis:

$$\mathbf{e}_1 = \mathbf{i} = (1,0,0); \mathbf{e}_2 = \mathbf{j} = (0,1,0), \mathbf{e}_3 = \mathbf{k} = (0,0,1).$$

And for any $\mathbf{a} = (a_1, a_2, a_3) \in E_3 W_n$ we get:

$$\mathbf{a} = a_1 \times_n \mathbf{i} +_n a_2 \times_n \mathbf{j} +_n a_3 \times_n \mathbf{k}$$

We name a scalar product of vectors

$$\mathbf{a} = (a_1, \ldots, a_m), \mathbf{b} = (b_1, \ldots, b_m) \in E_m W_n$$

the following sum:

$$(\mathbf{a}, \mathbf{b}) = (\ldots (a_1 \times_n b_1 +_n \ldots) +_n a_m \times_n b_m$$

We name a vector product of vectors

$$\mathbf{a} = (a_1, a_2, a_3), \mathbf{b} = (b_1, b_2, b_3) \in E_3 W_n$$

the following:

$$\mathbf{a} \times \mathbf{b} = (a_2 \times_n b_3 -_n a_3 \times_n b_2) \times_n \mathbf{i}_n$$
$$-_n (a_1 \times_n b_3 -_n a_3 \times_n b_1) \times_n \mathbf{j}_n$$
$$+_n (a_1 \times_n b_2 -_n a_2 \times_n b_1) \times_n \mathbf{k}$$

Let's consider the properties of the space $E_m W_n$.

DOI: 10.1201/9781003175902-3

Theorem 3.1 *If* $a +_n b \in E_m W_n$, *then* $b +_n a \in E_m W_n$ *and* $a +_n b = b +_n a$. *And inverse, if* $b +_n a \in E_m W_n$, *then* $a +_n b \in E_m W_n$ *and* $a +_n b = b +_n a$.

Theorem 3.2 *Addition associativity in* $E_m W_n$ *does not exist. But if*

$$a +_n b, (a +_n b) +_n c, b +_n c, a +_n (b +_n c) \in E_m W_n$$

then

$$(a +_n b) +_n c = a +_n (b +_n c)$$

It is important here the condition

$$a +_n b, (a +_n b) +_n c, b +_n c, a +_n (b +_n c) \in E_m W_n$$

We can see from the following example: For

$$m = 3, a = (50, 50, 50), b = (-50, -50, -50), c = (-50, -50, -50)$$

we have

$$a +_n b = 0, (a +_n b) +_n c = c$$

and $b +_n c$ does not belong to $E_3 W_n$.

Theorem 3.3 *The space* $E_m W_n$ *contains zero: for all* a

$$0 +_n a = a +_n 0 = a$$

where $0 = (0, 0, 0)$.

Theorem 3.4 *The space* $E_m W_n$ *contains inverse by addition element: for any* a, *there is* $(-a)$ *such that*

$$a +_n (-a) = 0.$$

There is no associativity of scalar multiplication. If

$$a = (a_1, \ldots, a_m) \in E_m W_n; \alpha, \beta \in W_n$$

then

$$\alpha \times_n (\beta \times_n a) = (\alpha \times_n (\beta \times_n a_1), \ldots, \alpha \times_n (\beta \times_n a_m))$$

and

$$(\alpha \times_n \beta) \times_n a = ((\alpha \times_n \beta) \times_n a_1, \ldots, (\alpha \times_n \beta) \times_n a_m)$$

we have to have

$$\alpha \times_n (\beta \times_n a_1), \ldots, \alpha \times_n (\beta \times_n a_m), (\alpha \times_n \beta) \times_n a_1, \ldots, (\alpha \times_n \beta) \times_n a_m \in W_n$$

If we take $\alpha = 1, \beta = 1, a = (1, 1, 1) \in E_3 W_n$, then

$$\alpha \times_n (\beta \times_n a) = (\alpha \times_n \beta) \times_n a$$

But if we take

$$\alpha = 0.01, \beta = 0.1, \mathbf{a} = (10, 10, 10) \in E_3 W_2$$

then

$$\alpha \times_n (\beta \times_n \mathbf{a}) = (0.01; 0.01; 0.01)$$

and

$$(\alpha \times_n \beta) \times_n \mathbf{a} = (0; 0; 0)$$

And we know

$$\delta_3 = \alpha \times_n (\beta \times_n \gamma) -_n (\alpha \times_n \beta) \times_n \gamma, (\alpha, \beta, \gamma \in W_n)$$

is a random variable in W_n, $\delta_3 = 0$ with probability $P < 1$. So, we get a theorem:

Theorem 3.5 *A probability of the equality*

$$\alpha \times_n (\beta \times_n \boldsymbol{a}) = (\alpha \times_n \beta) \times_n \boldsymbol{a}$$

is less than 1.

There is no distributivity of scalar multiplication. If

$$\mathbf{a} = (a_1, \ldots, a_m) \in E_m W_n, \alpha, \beta \in W_n$$

then

$$(\alpha +_n \beta) \times_n \mathbf{a} = ((\alpha +_n \beta) \times_n a_1, \ldots, (\alpha +_n \beta) \times_n a_m)$$

and

$$\alpha \times_n \mathbf{a} +_n \beta \times_n \mathbf{a} = (\alpha \times_n a_1 +_n \beta \times_n a_1, \ldots, \alpha \times_n a_m +_n \beta \times_n a_m)$$

We have to have

$$(\alpha +_n \beta) \times_n a_1, \ldots, (\alpha +_n \beta) \times_n a_m, \alpha \times_n a_1 +_n \beta \times_n a_1, \ldots$$

$$\ldots, \alpha \times_n a_m +_n \beta \times_n a_m \in W_n$$

If we take

$$\alpha = 1, \beta = 1, \mathbf{a} = (1, 1, 1) \in E_3 W_n$$

then

$$(\alpha +_n \beta) \times_n \mathbf{a} = \alpha \times_n \mathbf{a} +_n \beta \times_n \mathbf{a}$$

But if we take

$$\alpha = 0.03, \beta = 0.07, \mathbf{a} = (0.1, 0.1, 0.1) \in E_3 W_2$$

then

$$(\alpha +_n \beta) \times_n \mathbf{a} = (0.01, 0.01, 0.01)$$

and
$$\alpha \times_n \mathbf{a} +_n \beta \times_n \mathbf{a} = (0,0,0)$$

And we know
$$\delta_2 = (\alpha +_n \beta) \times_n \gamma -_n (\alpha \times_n \gamma +_n \beta \times_n \gamma), \alpha, \beta, \gamma \in W_n$$

is a random variable in W_n, and $\delta_2 = 0$ with probability $P < 1$. So, we get a theorem:

Theorem 3.6 *A probability of the equality*
$$(\alpha +_n \beta) \times_n \boldsymbol{a} = \alpha \times_n \boldsymbol{a} +_n \beta \times_n \boldsymbol{a}$$

is less than 1.

There is no distributivity of scalar multiplication to vector sum. If
$$\mathbf{a} = (a_1, \ldots, a_m), \mathbf{b} = (b_1, \ldots, b_m) \in E_m W_n, \alpha \in W_n$$

then
$$\alpha \times_n (\mathbf{a} +_n \mathbf{b}) = (\alpha \times_n (a_1 +_n b_1), \ldots, \alpha \times_n (a_m +_n b_m))$$

and
$$\alpha \times_n \mathbf{a} +_n \alpha \times_n \mathbf{b} = (\alpha \times_n a_1 +_n \alpha \times_n b_1, \ldots, \alpha \times_n a_m +_n \alpha \times_n b_m)$$

We have to have
$$\alpha \times_n (a_1 +_n b_1), \ldots, \alpha \times_n (a_m +_n b_m), \alpha \times_n a_1 +_n \alpha \times_n b_1, \ldots$$
$$\ldots, \alpha \times_n a_m +_n \alpha \times_n b_m \in W_n$$

If we take
$$\alpha = 1, \mathbf{a} = (1,1,1), \mathbf{b} = (1,1,1) \in E_3 W_n$$

then
$$\alpha \times_n (\mathbf{a} +_n \mathbf{b}) = \alpha \times_n \mathbf{a} +_n \alpha \times_n \mathbf{b}$$

But if we take
$$\alpha = 0.01, \mathbf{a} = (0.6, 0.6, 0.6), \mathbf{b} = (0.4, 0.4, 0.4) \in E_3 W_2$$

then
$$\alpha \times_n (\mathbf{a} +_n \mathbf{b}) = (0.01, 0.01, 0.01)$$

and
$$\alpha \times_n \mathbf{a} +_n \alpha \times_n \mathbf{b} = (0,0,0)$$

And we know
$$\delta_2 = (\alpha +_n \beta) \times_n \gamma -_n (\alpha \times_n \gamma +_n \beta \times_n \gamma), \alpha, \beta, \gamma \in W_n$$

is a random variable in W_n, and $\delta_2 = 0$ with probability $P < 1$. So, we get a theorem:

Theorem 3.7 *A probability of the equality*

$$\alpha \times_n (a +_n b) = \alpha \times_n a +_n \alpha \times_n b$$

is less than 1.

Also we get:

Theorem 3.8 *There is scalar multiplication unit:*

$$1 \times_n a = a$$

.

Now let's consider scalar product in $E_m W_n$.

Theorem 3.9 *Scalar product in $E_m W_n$ is commutative:*

$$(a, b) = (b, a)$$

.

Scalar product in $E_m W_n$ is not distributive. If

$$a = (a_1, \ldots, a_m), b = (b_1, \ldots, b_m),$$

$$c = (c_1, \ldots, c_m) \in E_m W_n$$

then

$$(\mathbf{a}, (\mathbf{b} +_n \mathbf{c})) = (\ldots (a_1 \times_n (b_1 +_n c_1) +_n \ldots) +_n a_m \times_n (b_m +_n c_m)$$

$$(\mathbf{a}, \mathbf{b}) +_n (\mathbf{a}, \mathbf{c}) = (a_1 \times_n b_1 +_n a_1 \times_n c_1) +_n \ldots) +_n (a_m \times_n b_m +_n a_m \times_n c_m)$$

We have to assume that all elements of these equalities are in W_n. If we take

$$a = (1, 1, 1), b = (1, 1, 1), c = (1, 1, 1) \in E_3 W_n$$

then

$$(\mathbf{a}, (\mathbf{b} +_n \mathbf{c})) = (\mathbf{a}, \mathbf{b}) +_n (\mathbf{a}, \mathbf{c})$$

But if we take

$$a = (0.01, 0.01, 0.01), b = (0.6, 0.6, 0.6), c = (0.4, 0.4, 0.4) \in E_3 W_2$$

then

$$(\mathbf{a}, (\mathbf{b} +_n \mathbf{c})) = 0.03$$

and

$$(\mathbf{a}, \mathbf{b}) +_n (\mathbf{a}, \mathbf{c}) = 0$$

And we know

$$\delta_2 = (\alpha +_n \beta) \times_n \gamma -_n (\alpha \times_n \gamma +_n \beta \times_n \gamma), \alpha, \beta, \gamma \in W_n$$

is a random variable in W_n, and $\delta_2 = 0$ with probability $P < 1$. So, we get a theorem:

Theorem 3.10 *A probability of*

$$(\mathbf{a}, (\mathbf{b} +_n \mathbf{c})) = (\mathbf{a}, \mathbf{b}) +_n (\mathbf{a}, \mathbf{c})$$

is less than 1.

Scalar multiplication on scalar product in $E_m W_n$ is not associative. If

$$\mathbf{a} = (a_1, \ldots, a_m)$$

$$\mathbf{b} = (b_1, \ldots, b_m) \in E_m W_n, \alpha \in W_n$$

then

$$\alpha \times_n (\mathbf{a}, \mathbf{b}) = \alpha \times_n (\ldots (a_1 \times_n b_1 +_n \ldots) +_n a_m \times_n b_m)$$

$$((\alpha \times_n \mathbf{a}), \mathbf{b}) = (\ldots (\alpha \times_n a_1) \times_n b_1 +_n \ldots) +_n (\alpha \times_n a_m) \times_n b_m$$

We have to assume that all elements of these equalities are in W_n. If we take

$$\alpha = 1, \mathbf{a} = (1, 1, 1), \mathbf{b} = (1, 1, 1) \in E_3 W_n$$

then

$$\alpha \times_n (\mathbf{a}, \mathbf{b}) = ((\alpha \times_n \mathbf{a}), \mathbf{b})$$

But if we take

$$\alpha = 0.01, \mathbf{a} = (0.1, 0.3, 0.6), \mathbf{b} = (1, 1, 1) \in E_3 W_2$$

then

$$\alpha \times_n (\mathbf{a}, \mathbf{b}) = 0.01$$

and

$$((\alpha \times_n \mathbf{a}), \mathbf{b}) = 0$$

And we know

$$\delta_3 = \alpha \times_n (\beta \times_n \gamma) -_n (\alpha \times_n \beta) \times_n \gamma, (\alpha, \beta, \gamma \in W_n)$$

is a random variable in W_n, and $\delta_3 = 0$ with probability $P < 1$. So, we get a theorem:

Theorem 3.11 *A probability of*

$$\alpha \times_n (\mathbf{a}, \mathbf{b}) = ((\alpha \times_n \mathbf{a}), \mathbf{b})$$

is less than 1.

We call vector's square of length

$$|\mathbf{a}|^2 = (\ldots (a_1 \times_n a_1 +_n \ldots) +_n a_m \times_n a_m)$$

Square of length $|a|^2$ of vector $\mathbf{a} = (a_1, \ldots, a_m)$ always exists, if

$$(\ldots (a_1 \times_n a_1 +_n \ldots) +_n a_m \times_n a_m) \in W_n$$

but length itself calculated as

$$\sqrt{|\mathbf{a}|^2} \in W_n$$

exists not always. For example, vectors of standard basis

$$\mathbf{i} = (1,0,0), \mathbf{j} = (0,1,0), \mathbf{k} = (0,0,1)$$

have length 1 for any W_n. But vector

$$\mathbf{a} = (0.7, 0.1, 0) \in E_3 W_2$$

does not have a length. Generally vector's length existence has some probability less than 1.

We call vectors $\mathbf{a}$ and $\mathbf{b}$ are perpendicular, if

$$(\mathbf{a}, \mathbf{b}) = 0$$

We call vectors $\mathbf{a}, \mathbf{b}$ are parallel ($\mathbf{a}\|\mathbf{b}$) if there are exist $\alpha \in W_n$, or $\beta \in W_n$ such that

$$\mathbf{b} = \alpha \times_n \mathbf{a}$$

or

$$\mathbf{a} = \beta \times_n \mathbf{b}$$

Two non-zero vectors in $E_m W_n$ may be perpendicular and parallel same time. For example, if

$$\mathbf{a} = (0.02, 0.04, 0.01) \in E_3 W_2, \alpha = 2, \mathbf{b} = \alpha \times_n \mathbf{a} = (0.04, 0.08, 0.02)$$

then

$$\mathbf{a}\|\mathbf{b}$$

and

$$(\mathbf{a}, \mathbf{b}) = 0$$

We can see also:

$$(\mathbf{i}, \mathbf{i}) = (\mathbf{j}, \mathbf{j}) = (\mathbf{k}, \mathbf{k}) = 1$$
$$|\mathbf{i}| = |\mathbf{j}| = |\mathbf{k}| = 1$$
$$(\mathbf{i}, \mathbf{j}) = (\mathbf{i}, \mathbf{k}) = (\mathbf{k}, \mathbf{j}) = 0$$

i.e. $\mathbf{i}, \mathbf{j}, \mathbf{k}$ is orthonormal basis in $E_3 W_n$.

Now let's consider vector product in $E_m W_n$.

For $\mathbf{a} = \mathbf{i} = (1,0,0), \mathbf{b} = \mathbf{j} = (0,1,0), \mathbf{c} = \mathbf{k} = (0,0,1)$ we have in $E_3 W_n$:

$$\mathbf{i} = \mathbf{j} \times \mathbf{k}, \mathbf{j} = \mathbf{k} \times \mathbf{i}, \mathbf{k} = \mathbf{i} \times \mathbf{j}$$

$$\mathbf{k} \times \mathbf{j} = -\mathbf{i}, \mathbf{i} \times \mathbf{k} = -\mathbf{j}, \mathbf{j} \times \mathbf{i} = -\mathbf{k}$$

$$\mathbf{i} \times \mathbf{i} = \mathbf{j} \times \mathbf{k} = \mathbf{k} \times \mathbf{k} = 0$$

Theorem 3.12

$$a \times a = 0$$

for any $a \in E_3 W_n$.

Theorem 3.13

$$a \times b = -(b \times a)$$

for any $a, b \in E_3 W_n$.

Vector product in $E_3 W_n$ is not distributive. For any

$$a = (a_1, a_2, a_3), b = (b_1, b_2, b_3), c = (c_1, c_2, c_3) \in E_3 W_n$$

$$a \times (b +_n c) = ((a_2 \times_n (b_3 +_n c_3) -_n a_3 \times_n (b_2 +_n c_2)) \times_n \mathbf{i} -_n$$
$$-_n (a_1 \times_n (b_3 +_n c_3) -_n (a_3 \times_n (b_1 +_n c_1)) \times_n \mathbf{j} +_n$$
$$+_n (a_1 \times_n (b_2 +_n c_2) -_n a_2 \times_n (b_1 +_n c_1)) \times_n \mathbf{k}$$

$$a \times b = (a_2 \times_n b_3 -_n a_3 \times_n b_2) \times_n \mathbf{i} -_n$$
$$-_n (a_1 \times_n b_3 -_n a_3 \times_n b_1) \times_n \mathbf{j} +_n$$
$$+_n (a_1 \times_n b_2 -_n a_2 \times_n b_1) \times_n \mathbf{k}$$

$$a \times c = (a_2 \times_n c_3 -_n a_3 \times_n c_2) \times_n \mathbf{i} -_n$$
$$-_n (a_1 \times_n c_3 -_n a_3 \times_n c_1) \times_n \mathbf{j} +_n$$
$$+_n (a_1 \times_n c_2 -_n a_2 \times_n c_1) \times_n \mathbf{k}$$

$$a \times b +_n a \times c = ((a_2 \times_n b_3 -_n a_3 \times_n b_2) +_n ((a2 \times_n c_3 -_n a_3 \times_n c_2)) \times_n \mathbf{i} -_n$$
$$-_n ((a_1 \times_n b_3 -_n a_3 \times_n b_1) +_n (a_1 \times_n c_3 -_n a_3 \times_n c_1)) \times_n \mathbf{j} +_n$$
$$+_n ((a_1 \times_n b_2 -_n a_2 \times_n b_1) +_n (a_1 \times_n c_2 -_n a_2 \times_n c_1)) \times_n \mathbf{k}$$

All elements of these equalities have to be in W_n. If we take

$$a = (1, 1, 1), b = (1, 1, 1), c = (1, 1, 1) \in E_3 W_n$$

then

$$a \times (b +_n c) =$$
$$= a \times b +_n a \times c$$

But if we take

$$a = (0.01, 0.02, 0.03), b = (0.6, 0.6, 0.6),$$
$$c = (0.4, 0.4, 0.4) \in E_3 W_2$$

then

$$a \times (b +_n c) = (-0.01, 0.02, -0.01)$$

and

$$a \times b +_n a \times c = (0, 0, 0)$$

And we know

$$\delta_2 = (\alpha +_n \beta) \times_n \gamma -_n (\alpha \times_n \gamma +_n \beta \times_n \gamma)$$

$$\alpha, \beta, \gamma \in W_n$$

is a random variable in W_n, and $\delta_2 = 0$ with probability $P < 1$. So, we get a theorem:

Theorem 3.14 *A probability of equality*

$$\boldsymbol{a} \times (\boldsymbol{b} +_n \boldsymbol{c}) = \boldsymbol{a} \times \boldsymbol{b} +_n \boldsymbol{a} \times \boldsymbol{c}$$

is less than 1.

Scalar multiplication on vector product in $E_3 W_n$ is not associative. For any

$$\mathbf{a} = (a_1, a_2, a_3), \mathbf{b} = (b_1, b_2, b_3) \in E_3 W_n$$

and for any scalar $\alpha \in W_n$

$$(\alpha \times_n \mathbf{a}) \times \mathbf{b} = ((\alpha \times_n a_2) \times_n b_3 -_n (\alpha \times_n a_3) \times_n b_2) \times_n \mathbf{i} -_n$$
$$-_n ((\alpha \times_n a_1) \times_n b_3 -_n (\alpha \times_n a_3) \times_n b_1) \times_n \mathbf{j} +_n$$
$$+_n ((\alpha \times_n a_1) \times_n b_2 -_n (\alpha \times_n a_2) \times_n b_1) \times_n \mathbf{k}$$

$$\mathbf{a} \times (\alpha \times_n \mathbf{b}) = ((a_2 \times_n (\alpha \times_n b_3) -_n a_3 \times_n (\alpha \times_n b_2)) \times_n \mathbf{i} -_n$$
$$-_n (a_1 \times_n (\alpha \times_n b_3) -_n a_3 \times_n (\alpha \times_n b_1)) \times_n \mathbf{j} +_n$$
$$+_n (a_1 \times_n (\alpha \times_n b_2) -_n a_2 \times_n (\alpha \times_n b_1)) \times_n \mathbf{k}$$

$$\alpha \times_n (\mathbf{a} \times \mathbf{b}) = (\alpha \times_n (a_2 \times_n b_3 -_n a_3 \times_n b_2)) \times_n \mathbf{i} -_n$$
$$-_n (\alpha \times_n (a_1 \times_n b_3 -_n a_3 \times_n b_1)) \times_n \mathbf{j} +_n$$
$$+_n (\alpha \times_n (a_1 \times_n b_2 -_n a_2 \times_n b_1)) \times_n \mathbf{k}$$

All elements of these equalities have to be in W_n. If we take

$$\alpha = 1, \mathbf{a} = (1, 1, 1), \mathbf{b} = (1; 1; 1) \in E_3 W_n$$

then

$$\alpha \times_n (\mathbf{a} \times \mathbf{b}) = (\alpha \times_n \mathbf{a}) \times \mathbf{b}$$

But if we take

$$\alpha = 0.01, \mathbf{a} = (0.1, 0.3, 0.6), \mathbf{b} = (10; 10; 10) \in E_3 W_2$$

then

$$\alpha \times_n (\mathbf{a} \times \mathbf{b}) = (-0.03, 0, 05, -0.02)$$

and

$$(\alpha \times_n \mathbf{a}) \times \mathbf{b} = (0, 0, 0)$$

And we know

$$\delta_2 = (\alpha +_n \beta) \times_n \gamma -_n (\alpha \times_n \gamma +_n \beta \times_n \gamma), \alpha, \beta, \gamma \in W_n$$
$$\delta_3 = \alpha \times_n (\beta \times_n \gamma) -_n (\alpha \times_n \beta) \times_n \gamma, (\alpha, \beta, \gamma \in W_n)$$

are the random variables in W_n, and $\delta_2 = \delta_3 = 0$ with probability $P < 1$. So, we get a theorem:

Theorem 3.15 *A probability of equality*

$$\alpha \times_n (\boldsymbol{a} \times \boldsymbol{b}) = (\alpha \times_n \boldsymbol{a}) \times \boldsymbol{b}$$

is less than 1.

For any

$$\mathbf{a} = (a_1, a_2, a_3), \mathbf{b} = (b_1, b_2, b_3), \mathbf{c} = (c_1, c_2, c_3) \in E_3 W_n$$

$$\mathbf{a} \times (\mathbf{b} \times \mathbf{c}) = \mathbf{a} \times ((b_2 \times_n c_3 -_n b_3 \times_n c_2) \times_n \mathbf{i} -_n$$
$$-_n (b_1 \times_n c_3 -_n b_3 \times_n c_1) \times_n \mathbf{j} +_n$$
$$+_n (b_1 \times_n c_2 -_n b_2 \times_n c_1) \times_n \mathbf{k} =$$
$$= (a_2 \times_n (b_1 \times_n c_2 -_n b_2 \times_n c_1) -_n a_3 \times_n (-b_1 \times_n c_3 +_n b_3 \times_n c_1)) \times_n \mathbf{i} -_n$$
$$-_n (a_1 \times_n (b_1 \times_n c_2 -_n b_2 \times_n c_1) -_n a_3 \times_n (b_2 \times_n c_3 -_n b_3 \times_n c_2)) \times_n \mathbf{j} +_n$$
$$+_n (a_1 \times_n (-b_1 \times_n c_3 +_n b_3 \times_n c_1) -_n a_2 \times_n (b_1 \times_n c_2 -_n b_2 \times_n c_1)) \times_n \mathbf{k}$$
$$(\mathbf{a}, \mathbf{c}) \times_n \mathbf{b} = ((a_1 \times_n c_1 +_n a_2 \times_n c_2) +_n a_3 \times_n c_3) \times_n b_1) \times_n \mathbf{i} +_n$$
$$+_n ((a_1 \times_n c_1 +_n a_2 \times_n c_2) +_n a_3 \times_n c_3) \times_n b_2) \times_n \mathbf{j} +_n$$
$$+_n ((a_1 \times_n c_1 +_n a_2 \times_n c_2) +_n a_3 \times_n c_3) \times_n b_3) \times_n \mathbf{k}$$
$$(\mathbf{a}, \mathbf{b}) \times_n \mathbf{c} = ((a_1 \times_n b_1 +_n a_2 \times_n b_2) +_n a_3 \times_n b_3) \times_n c_1) \times_n \mathbf{i} +_n$$
$$+_n ((a_1 \times_n b_1 +_n a_2 \times_n b_2) +_n a_3 \times_n b_3) \times_n c_2) \times_n \mathbf{j} +_n$$
$$+_n ((a_1 \times_n b_1 +_n a_2 \times_n b_2) +_n a_3 \times_n b_3) \times_n c_3) \times_n \mathbf{k}$$

All elements of these equalities have to be in W_n. If we take

$$\mathbf{a} = (1, 1, 1), \mathbf{b} = (1; 1; 1), \mathbf{c} = (1; 1; 1) \in E_3 W_n$$

then

$$\mathbf{a} \times (\mathbf{b} \times \mathbf{c}) = (\mathbf{a}, \mathbf{c}) \times_n \mathbf{b} -_n (\mathbf{a}, \mathbf{b}) \times_n \mathbf{c}$$

But if we take

$$\mathbf{a} = (0.01, 0.01, 0.01), \mathbf{b} = (0.6; 0.6; 0.6), \mathbf{c} = (0.4; 0.9; -0.9) \in E_3 W_2$$

then

$$\mathbf{a} \times (\mathbf{b} \times \mathbf{c}) = (0; -0.01; 0.01)$$

and

$$(\mathbf{a}, \mathbf{c}) \times_n \mathbf{b} -_n (\mathbf{a}, \mathbf{b}) \times_n \mathbf{c} = (0, 0, 0)$$

And we know

$$\delta_2 = (\alpha +_n \beta) \times_n \gamma -_n (\alpha \times_n \gamma +_n \beta \times_n \gamma), \alpha, \beta, \gamma \in W_n$$

$$\delta_3 = \alpha \times_n (\beta \times_n \gamma) -_n (\alpha \times_n \beta) \times_n \gamma, (\alpha, \beta, \gamma \in W_n)$$

are the random variables in W_n, and $\delta_2 = \delta_3 = 0$ with probability $P < 1$. So, we get a theorem:

Theorem 3.16 *A probability of equality*

$$\boldsymbol{a} \times (\boldsymbol{b} \times \boldsymbol{c}) = (\boldsymbol{a}, \boldsymbol{c}) \times_n \boldsymbol{b} -_n (\boldsymbol{a}, \boldsymbol{b}) \times_n \boldsymbol{c}$$

is less than 1.

We named vectors $\mathbf{a}, \mathbf{b}$ are parallel $(\mathbf{a}\|\mathbf{b})$, if there are $\alpha, \beta \in W_n$ such that $\mathbf{b} = \alpha \times_n \mathbf{a}$ or $\mathbf{a} = \beta \times_n \mathbf{b}$. For any

$$\alpha \in W_n, \mathbf{a} = (a_1, a_2, a_3), \mathbf{b} = \alpha \times_n \mathbf{a} = (\alpha \times_n a_1, \alpha \times_n a_2, \alpha \times_n a_3) \in E_3 W_n$$

$$\mathbf{a} \times \mathbf{b} = ((a_2 \times_n (\alpha \times_n a_3) -_n a_3 \times_n (\alpha \times_n a_2)) \times_n \mathbf{i} -_n$$

$$-_n (a_1 \times_n (\alpha \times_n a_3) -_n a_3 \times_n (\alpha \times_n a_1)) \times_n \mathbf{j} +_n$$

$$+_n (a_1 \times_n (\alpha \times_n a_2) -_n a_2 \times_n (\alpha \times_n a_1)) \times_n \mathbf{k}$$

And $\mathbf{a} \times \mathbf{b} = \mathbf{0}$ also with probability less than 1, because we know that

$$\delta_2 = (\alpha +_n \beta) \times_n \gamma -_n (\alpha \times_n \gamma +_n \beta \times_n \gamma), \alpha, \beta, \gamma \in W_n$$

$$\delta_3 = \alpha \times_n (\beta \times_n \gamma) -_n (\alpha \times_n \beta) \times_n \gamma, (\alpha, \beta, \gamma \in W_n)$$

are the random variables in W_n, and $\delta_2 = \delta_3 = 0$ with probability $P < 1$. Difference between $\mathbf{a} \times \mathbf{b}$ and $\mathbf{0}$ is changing with growing n. Below we give some examples. Let's

$$\mathbf{b} = \alpha \times_n \mathbf{a} = (\alpha \times_n a_1, \alpha \times_n a_2, \alpha \times_n a_3)$$

with $\alpha, \alpha \times_n a_1, \alpha \times_n a_2, \alpha \times_n a_3 \in W_n$

Let's take

$$n = 4, \mathbf{a} = (3.1549, 2.9807, 1.7362), \alpha = 4.4697$$

Then

$$\mathbf{b} = (14.1002, 13.3211, 7.759)$$

$$\mathbf{a} \times \mathbf{b} = -0.0018\mathbf{i} +_n 0.0031\mathbf{j} -_n 0.0019\mathbf{k}$$

$$(\mathbf{a}, \mathbf{a} \times \mathbf{b}) = -0.0055 +_n 0.0089 -_n 0.014 = -0.0106$$

$$(\mathbf{b}, \mathbf{a} \times \mathbf{b}) = -0.0253 +_n 0.0412 -_n 0.014 = 0.0019$$

Let's take

$$n = 6, \mathbf{a} = (3.154932, 2.980749, 1.736284), \alpha = 4.469731$$

Then

$$\mathbf{b} = (14.101681, 13.323129, 7.760707)$$

$$\mathbf{a} \times \mathbf{b} = -0.000021\mathbf{i} +_n 0.000020\mathbf{j} -_n 0.000003\mathbf{k}$$

$$(\mathbf{a}, \mathbf{a} \times \mathbf{b}) = -0.000065 +_n 0.000058 -_n 0.000003 = -0.00001$$

$$(\mathbf{b}, \mathbf{a} \times \mathbf{b}) = -0.000296 +_n 0.000266 -_n 0.000021 = -0.000051$$

Let's take

$$n = 8, \mathbf{a} = (3.15493269, 2.98074951, 1.73628439), \alpha = 4.46973129$$

Then

$$\mathbf{b} = (14.10170118, 13.32314907, 7.76072442)$$

$$\mathbf{a} \times \mathbf{b} = -0.00000021\mathbf{i} +_n 0.00000051\mathbf{j} -_n 0.00000036\mathbf{k}$$

$$(\mathbf{a}, \mathbf{a} \times \mathbf{b}) = -0.00000065 +_n 0.00000147 -_n 0.00000057 = 0.00000025$$

$$(\mathbf{b}, \mathbf{a} \times \mathbf{b}) = -0.00000296 +_n 0.00000678 -_n 0.00000273 = 0.00000109$$

If we take $\mathbf{a} = \mathbf{i}, \mathbf{b} = \mathbf{j}$, then $\mathbf{a}$ and $\mathbf{b}$ are not parallel, and $\mathbf{a} \times \mathbf{b} \neq \mathbf{0}$. But if we take $\mathbf{a} = (0.01, 0.05, 0.07), \mathbf{b} = (0.07, 0.02, 0.06), \mathbf{a}, \mathbf{b} \in E_3 W_2$ then $\mathbf{a}$ and $\mathbf{b}$ are not parallel, and $\mathbf{a} \times \mathbf{b} = \mathbf{0}$.

For any $\mathbf{a} = (a_1, a_2, a_3), \mathbf{b} = (b_1, b_2, b_3) \in E_3 W_n$

$$(\mathbf{a}, \mathbf{a} \times \mathbf{b}) = (a_1 \times_n (a_2 \times_n b_3 -_n a_3 \times_n b_2) -_n a_2 \times_n (a_1 \times_n b_3 -_n a_3 \times_n b_1)) +_n$$

$$+_n a_3 \times_n (a_1 \times_n b_2 -_n a_2 \times_n b_1)$$

And $(\mathbf{a}, \mathbf{a} \times \mathbf{b}) = \mathbf{0}$ also with probability less than 1, because we know that

$$\delta_2 = (\alpha +_n \beta) \times_n \gamma -_n (\alpha \times_n \gamma +_n \beta \times_n \gamma), \alpha, \beta, \gamma \in W_n$$

$$\delta_3 = \alpha \times_n (\beta \times_n \gamma) -_n (\alpha \times_n \beta) \times_n \gamma, (\alpha, \beta, \gamma \in W_n)$$

are the random variables in W_n, and $\delta_2 = \delta_3 = 0$ with probability $P < 1$. So, we get a theorem:

Theorem 3.17 *The probabilities of correctness of the following statements: "If $\boldsymbol{a}$ and $\boldsymbol{b}$ are parallel ($\boldsymbol{a} \| \boldsymbol{b}$), then $\boldsymbol{a} \times \boldsymbol{b} = \boldsymbol{0}$" and "If $\boldsymbol{a}$ and $\boldsymbol{b}$ are not parallel, then $\boldsymbol{a} \times \boldsymbol{b} \neq 0$ and $(\boldsymbol{a}, \boldsymbol{a} \times \boldsymbol{b}) = (\boldsymbol{b}, \boldsymbol{a} \times \boldsymbol{b}) = 0$" are less than 1.*

For any $\mathbf{a} = (a_1, a_2, a_3), \mathbf{b} = (b_1, b_2, b_3), \mathbf{c} = (c_1, c_2, c_3) \in E_3 W_n$

$$(\mathbf{a}, \mathbf{b} \times \mathbf{c}) =$$

$$= (a_1 \times_n (b_2 \times_n c_3 -_n b_3 \times_n c_2) -_n a_2 \times_n (b_1 \times_n c_3 -_n b_3 \times_n c_1)) +_n$$

$$+_n a_3 \times_n (b_1 \times_n c_2 -_n b_2 \times_n c_1)$$

$$(\mathbf{a} \times \mathbf{b}, \mathbf{c}) =$$

$$= ((a_2 \times_n b_3 -_n a_3 \times_n b_2) \times_n c_1 -_n (a_1 \times_n b_3 -_n a_3 \times_n b_1) \times_n c_2) +_n$$

$$+_n (a_1 \times_n b_2 -_n a_2 \times_n b_1) \times_n c_3$$

All elements of these equalities have to be in W_n. If we take

$$\mathbf{a} = (1, 1, 1), \mathbf{b} = (1; 1; 1)$$

,

$$\mathbf{c} = (1; 1; 1) \in E_3 W_n$$

then

$$(\mathbf{a}, \mathbf{b} \times \mathbf{c}) =$$
$$= (\mathbf{a} \times \mathbf{b}, \mathbf{c})$$

But if we take

$$\mathbf{a} = (0.01, 0.01, 0.01), \mathbf{b} = (0.6; 0.6; 0.6)$$
$$\mathbf{c} = (0.4; 0.9; -0.9) \in E_3 W_2$$

then

$$(\mathbf{a}, \mathbf{b} \times \mathbf{c}) = -0.01$$

and

$$(\mathbf{a} \times \mathbf{b}, \mathbf{c}) = 0$$

And

$$(\mathbf{a}, \mathbf{b} \times \mathbf{c}) = (\mathbf{a} \times \mathbf{b}, \mathbf{c})$$

with probability less than 1, because we know that

$$\delta_2 = (\alpha +_n \beta) \times_n \gamma -_n (\alpha \times_n \gamma +_n \beta \times_n \gamma)$$

$$\alpha, \beta, \gamma \in W_n$$

$$\delta_3 = \alpha \times_n (\beta \times_n \gamma) -_n (\alpha \times_n \beta) \times_n \gamma, (\alpha, \beta, \gamma \in W_n)$$

are the random variables in W_n, and $\delta_2 = \delta_3 = 0$ with probability $P < 1$.

So, we get a theorem:

Theorem 3.18 *Equality*

$$(a, b \times c) = (a \times b, c)$$

is correct in $E_3 W_n$ with probability less than 1.

4

Observability and Mathematical Analysis (Calculus)

We call "right derivative of function $f(x)$ at point x" and note it as "$\frac{df}{dx}(x)_+$" the following expression

$$\frac{df}{dx}(x)_+ = (f(x +_n \Delta x) -_n f(x)) \times_n \left(\frac{1}{\Delta x}\right)$$

if

$$f, x, \Delta x, (f(x +_n \Delta x) -_n f(x)) \times_n \left(\frac{1}{\Delta x}\right) \in W_n$$

and Δx – positive number with existing $\frac{1}{\Delta x} \in W_n$. The value of Δx is a derivative's parameter. The main value of parameter Δx is $\Delta x_{min} = 0.\underbrace{0...0}_{n-1}2$.

In this case $\frac{1}{\Delta x_{min}} = 5\underbrace{0...0}_{n-1}.\underbrace{*...*}_{n}$, where any $*$ is an arbitrary digit $\in (0,1,2,3,4,5,6,7,8,9)$.

Example 1: $n = 2$, $\Delta x = 0.02$, $\frac{1}{\Delta x} = 50.00, 50.01, ..., 50.09, 50.10, 50.11, ..., 50.99$,

$$f(x +_n \Delta x) -_n f(x) = 0.09$$

$$\frac{df}{dx}(x)_+ = 4.50.$$

We name function $f(x)$ the right differentiable function at point x, if $\frac{df}{dx}(x)_+$ exists.

Example 2: $n = 2$, $\Delta x = 0.02$, $\frac{1}{\Delta x} = 50.00, 50.01, ..., 50.09, 50.10, 50.11, ..., 50.99$,

$$f(x +_n \Delta x) -_n f(x) = 2.00$$

$\frac{df}{dx}(x)_+$ does not exist.

We call "left derivative of function $f(x)$ at point x" and note it as "$\frac{df}{dx}(x)_-$" the following expression

$$\frac{df}{dx}(x)_- = (f(x +_n \Delta x) -_n f(x)) \times_n \left(\frac{1}{\Delta x}\right)$$

if

$$f, x, \Delta x, (f(x +_n \Delta x) -_n f(x)) \times_n \left(\frac{1}{\Delta x}\right) \in W_n$$

DOI: 10.1201/9781003175902-4

and Δx – negative number with existing $\frac{1}{\Delta x} \in W_n$. The value of Δx is a derivative's parameter. The main value of parameter Δx in this case is $\Delta x_{max} = -0.\underbrace{0...0}_{n-1}2$. In this case $\frac{1}{\Delta x_{max}} = -5\underbrace{0...0}_{n-1}.\underbrace{*...*}_{n}$, where any $*$ is an arbitrary digit $\in (0,1,2,3,4,5,6,7,8,9)$.

Example 3: $n = 2$, $\Delta x = -0.02$, $\frac{1}{\Delta x} = -50.00, -50.01, ..., -50.99$,

$$f(x +_n \Delta x) -_n f(x) = -0.09$$

$$\frac{df}{dx}(x)_- = 4.50$$

We name function $f(x)$ the left differentiable function at point x, if $\frac{df}{dx}(x)_-$ exists.

Example 4: $n = 2$, $\Delta x = -0.02$, $\frac{1}{\Delta x} = -50.00, -50.01, ..., -50.99$,

$$f(x +_n \Delta x) -_n f(x) = -2$$

$\frac{df}{dx}(x)_-$ does not exist.

We name function $f(x)$ the differentiable function at point x, if $\frac{df}{dx}(x)_+$ and $\frac{df}{dx}(x)_-$ exist, and

$$\frac{df}{dx}(x)_+ = \frac{df}{dx}(x)_-$$

In this case we call "derivative of function $f(x)$ at point x " and note it as "$\frac{df}{dx}(x)$", if

$$\frac{df}{dx}(x) = \frac{df}{dx}(x)_+ = \frac{df}{dx}(x)_-$$

Let f be the function of several variables $x, y, z, ...$ We call "right partial derivative of function $f(x, y, z, ...)$ by variable x at point $(x_0, y, z, ..)$ " and note it as

$$(\partial f/\partial x)_+(x_0, y, z, ..)$$

the following expression

$$(\partial f/\partial x)_+(x_0, y, z, ..) = (f(x_0 +_n \Delta x, y, z, ...) -_n f(x_0, y, z, ...)) \times_n \left(\frac{1}{\Delta x}\right)$$

if variables $y, z, ...$ are fixed,

$$f, x_0, y, z, ..., \Delta x, (f(x_0 +_n \Delta x, y, z, ...) -_n f(x_0, y, z, ...)) \times_n \left(\frac{1}{\Delta x}\right) \in W_n$$

and Δx – positive number with existing $\frac{1}{\Delta x} \in W_n$.

We call "left partial derivative of function $f(x, y, z, ...)$ by variable x at point $(x_0, y, z, ..)$ " and note it as

$$(\partial f/\partial x)_-(x_0, y, z, ..)$$

the following expression

$$(\partial f/\partial x)_-(x_0, y, z, ..) = (f(x_0 +_n \Delta x, y, z, ...) -_n f(x_0, y, z, ...)) \times_n \left(\frac{1}{\Delta x}\right)$$

if variables $y, z, ...$ are fixed,

$$f, x_0, y, z, ..., \Delta x, (f(x_0 +_n \Delta x, y, z, ...) -_n f(x_0, y, z, ...)) \times_n \left(\frac{1}{\Delta x}\right) \in W_n$$

and Δx – negative number with existing $\frac{1}{\Delta x} \in W_n$.

We call " partial derivative of function $f(x, y, z, ...)$ by variable x at point $(x_0, y, z, ..)$ " and note it as

$$\partial f/\partial x(x_0, y, z, ..)$$

if

$$(\partial f/\partial x)_+(x_0, y, z, ..), (\partial f/\partial x)_-(x_0, y, z, ..)$$

exist and

$$(\partial f/\partial x)_+(x_0, y, z, ..) = (\partial f/\partial x)_-(x_0, y, z, ..)$$

On this case

$$\partial f/\partial x(x_0, y, z, ..) = (\partial f/\partial x)_+(x_0, y, z, ..) = (\partial f/\partial x)_-(x_0, y, z, ..)$$

Example 5:
Let's

$$n = 2, f(x) = x^2$$

and first take $x = 3$. We know that

$$\Delta x = 0.02, \frac{1}{\Delta x} = 50.00, 50.01, ..., 50.09, 50.10, 50.11, ..., 50.99$$

We get

$$\frac{df}{dx}(3)_+ = 6.00, 6.01, 6.02, 6.03, 6.04, 6.05, 6.06, 6.07, 6.08, 6.09$$

Let's take now

$$\Delta x = -0.02, \frac{1}{\Delta x} = -50.00, -50.01, ..., -50.09, -50.10, -50.11, ..., -50.99$$

We get

$$\frac{df}{dx}(3)_- = 13.50, 13.52, 13.54, 13.56, 13.58, 13.60, 13.62, 13.64, 13.66, 13.68$$

That means $\frac{df}{dx}(3)_+$ and $\frac{df}{dx}(3)_-$ are two different random variables. So, function $f(x)$ is right differentiable function at point $x = 3$ and is left differentiable function at same point, but is not differentiable function at point $x = 3$.

Let's take now $x = 3.24$. We get

$$\frac{df}{dx}(3.24)_+ = 6.00, 6.01, 6.02, 6.03, 6.04, 6.05, 6.06, 6.07, 6.08, 6.09$$

And

$$\frac{df}{dx}(3.24)_- = 6.00, 6.01, 6.02, 6.03, 6.04, 6.05, 6.06, 6.07, 6.08, 6.09$$

That means $\frac{df}{dx}(3.24)_+$ and $\frac{df}{dx}(3.24)_-$ are two equal random variables. So, function $f(x)$ is right differentiable function at point $x = 3.24$ and is left differentiable function at same point, and is differentiable function at point $x = 3.24$.

Let's formulate now the Integral definition.

We call " integral of function $f(x)$ on segment $[a, b]$ " and note it as "$\int_a^b f(x) \times_n \Delta x$" the following expression

$$\int_a^b f(x) \times_n \Delta x = \sum_{i=1}^{k} {}^n f(x_i) \times_n \Delta x$$

if

$$x, x_i, f(x), f(x_i), a, b, \Delta x, k, \sum_{i=1}^{k} {}^n f(x_i) \times_n \Delta x \in W_n$$

Also we assume that segment $[a, b]$ is divided on parts

$$[x_i, x_{i+_n 1}], i = 1, \ldots, k$$

where
$$x_1 = a, x_{k+_n 1} = b, x_{i+_n 1} -_n x_i = \Delta x, k \times_n \Delta x = b -_n a$$

As we did it above in derivative definition, the value of Δx is an integral parameter. The main value of parameter Δx is $\Delta x_{min} = 0.\underbrace{0...0}_{n-1}2$. So, for given n we have limiting on numbers a, b, k. For vector $\mathbf{a} = (a_1, a_2, a_3) \in E_3 W_n$ $a_1 = a_1(x), a_2 = a_2(x), a_3 = a_3(x), x \in W_n$ we call

$$\int_a^b \mathbf{a} \times_n \Delta x = (\int_a^b a_1(x) \times_n \Delta x, \int_a^b a_2(x) \times_n \Delta x, \int_a^b a_3(x) \times_n \Delta x) \in E_3 W_n$$

Example 6:
Let's consider function

$$y = f(x) = x; x, y \in W_n$$

For any

$$\Delta x \in W_n$$

with existing inverse

$$\frac{1}{\Delta x} \in W_n$$

we get

$$\frac{df}{dx}(x)_- = \frac{df}{dx}(x)_+ = \frac{df}{dx}(x) = 1$$

Example 7:
For any

$$\alpha = const, y = f(x) = \alpha \times_n x; x, y, \alpha \in W_n, \Delta x \in W_n, (x +_n \Delta x) \in W_n$$

we get

$$\frac{df}{dx}(x) = (f(x +_n \Delta x) -_n f(x)) \times_n \left(\frac{1}{\Delta x}\right) =$$

$$= (\alpha \times_n (x +_n \Delta x) -_n \alpha \times_n x) \times_n \left(\frac{1}{\Delta x}\right) =$$

$$= (\alpha \times_n \Delta x +_n \sigma_1) \times_n \left(\frac{1}{\Delta x}\right) =$$

$$= (\alpha \times_n \Delta x) \times_n \left(\frac{1}{\Delta x}\right) +_n \sigma_1 \times_n \left(\frac{1}{\Delta x}\right) +_n \sigma_2 =$$

$$= \alpha \times_n \left(\Delta x \times_n \frac{1}{\Delta x}\right) +_n \sigma_3 +_n \sigma_1 \times_n \left(\frac{1}{\Delta x}\right) +_n \sigma_2 =$$

$$= \alpha +_n \sigma_1 \times_n \left(\frac{1}{\Delta x}\right) +_n \sigma_2 +_n \sigma_3 = \alpha +_n \sigma_4$$

where $\sigma_1, \sigma_2, \sigma_3, \sigma_4$ are the random variables depend on n and m.
Let's consider function

$$y = f(x) = \alpha \times_n x; \alpha, x, y \in W_n$$

with

$$n = 2; \Delta x = 0.02; \alpha = 1.53; x = 4.23.$$

Direct calculations show

$$\frac{1}{\Delta x} = \frac{1}{0.02} = 50.00, 50.01, \ldots, 50.99$$

We get

$$\frac{df}{dx}(4.23)_+ = (1.53 \times_2 4.25 -_2 1.53 \times_2 4.23) \times_2 \frac{1}{\Delta x} =$$

$$= (6.47 -_2 6.45) \times_2 \frac{1}{\Delta x} = 0.02 \times_2 50 = 1$$

$$\frac{df}{dx}(4.23)_- = (1.53 \times_2 4.21 -_2 1.53 \times_2 4.23) \times_2 \frac{1}{-\Delta x} =$$

$$= (6.43 -_2 6.45) \times_2 \frac{1}{-\Delta x} = -0.02 \times_2 (-50) = 1$$

So,

$$\frac{df}{dx}(4.23)_- = \frac{df}{dx}(4.23)_+ = \frac{df}{dx}(4.23) = 1$$

And we have

$$1 = \alpha +_2 \sigma_4 = 1.53 +_2 \sigma_4$$

i.e.

$$\sigma_4 = -0.53$$

Let's take now

$$x = 4.25$$

We get

$$\frac{df}{dx}(4.25)_+ = (1.53 \times_2 4.27 -_2 1.53 \times_2 4.25) \times_2 \frac{1}{\Delta x} =$$

$$= (6.49 -_2 6.47) \times_2 \frac{1}{\Delta x} = 0.02 \times_2 50 = 1$$

$$\frac{df}{dx}(4.25)_- = (1.53 \times_2 4.23 -_2 1.53 \times_2 4.25) \times_2 \frac{1}{-\Delta x} =$$

$$= (6.45 -_2 6.47) \times_2 \frac{1}{-\Delta x} = -0.02 \times_2 (-50) = 1$$

So,

$$\frac{df}{dx}(4.25)_- = \frac{df}{dx}(4.25)_+ = \frac{df}{dx}(4.25) = 1$$

So, we have again

$$1 = \alpha +_2 \sigma_4 = 1.53 +_2 \sigma_4$$

i.e.

$$\sigma_4 = -0.53$$

Let's take now

$$x = 4.27$$

We get

$$\frac{df}{dx}(4.27)_+ = (1.53 \times_2 4.29 -_2 1.53 \times_2 4.27) \times_2 \frac{1}{\Delta x} =$$

$$= (6.51 -_2 6.49) \times_2 \frac{1}{\Delta x} = 0.02 \times_2 50 = 1$$

$$\frac{df}{dx}(4.27)_- = (1.53 \times_2 4.25 -_2 1.53 \times_2 4.27) \times_2 \frac{1}{-\Delta x} =$$

$$= (6.47 -_2 6.49) \times_2 \frac{1}{-\Delta x} = -0.02 \times_2 (-50) = 1$$

So,

$$\frac{df}{dx}(4.27)_- = \frac{df}{dx}(4.27)_+ = \frac{df}{dx}(4.27) = 1$$

So, we have again
$$1 = \alpha +_2 \sigma_4 = 1.53 +_2 \sigma_4$$

i.e.
$$\sigma_4 = -0.53$$

And we can calculate second derivative:

$$\frac{d^2 f}{dx^2}(4.25)_+ = \frac{d}{dx_+}\left(\frac{df}{dx}(4.25)\right) = \left(\frac{df}{dx}(4.27) -_2 \frac{df}{dx}(4.25)\right) \times_2 \frac{1}{\Delta x} = 0$$

$$\frac{d^2 f}{dx^2}(4.25)_- = \frac{d}{dx_-}\left(\frac{df}{dx}(4.25)\right) = \left(\frac{df}{dx}(4.23) -_2 \frac{df}{dx}(4.25)\right) \times_2 \frac{1}{\Delta x} = 0$$

So,

$$\frac{d^2 f}{dx^2}(4.25)_+ = \frac{d^2 f}{dx^2}(4.25)_- = \frac{d^2 f}{dx^2}(4.25)$$

Example 8:
Generally the function

$$y = f(x) = 1.53 \times_n x; x, y \in W_n$$

on the segment

$$[4.20, 4.29]$$

has first derivative at points

$$x = 4.22, 4.23, 4.24, 4.25, 4.26, 4.27$$

and does not have first derivative at points

$$x = 4.20, 4.21, 4.28, 4.29$$

And this function has second derivative at points

$$x = 4.24, 4.25$$

and does not have second derivative at points

$$x = 4.20, 4.21, 4.22, 4.23, 4.26, 4.27, 4.28, 4.29$$

Example 9:
Let's take now

$$x = 1.79$$

We get

$$\frac{df}{dx_+}(1.79) = (1.53 \times_2 1.81 -_2 1.53 \times_2 1.79) \times_2 \frac{1}{\Delta x} =$$

$$= (2.74 -_2 2.67) \times_2 \frac{1}{\Delta x} = 0.07 \times_2 50.0 = 3.5$$

$$\frac{df}{dx}_-(1.79) = (1.53 \times_2 1.77 -_2 1.53 \times_2 1.79) \times_2 \frac{1}{-\Delta x} =$$

$$= (2.65 -_2 2.67) \times_2 \frac{1}{\Delta x} = -0.02 \times_2 (-50.0) = 1$$

So, $\frac{df}{dx}(1.79)$ does not exist.

Example 10:
Let's take

$$\alpha = 3, x = 4.23$$

We get

$$\frac{df}{dx}(4.23)_+ = (3 \times_2 4.25 -_2 3 \times_2 4.23) \times_2 \frac{1}{\Delta x} =$$

$$= (12.75 -_2 12.69) \times_2 \frac{1}{\Delta x} = 0.06 \times_2 50 = 3 = \alpha$$

$$\frac{df}{dx}(4.23)_- = (3 \times_2 4.21 -_2 3 \times_2 4.23) \times_2 \frac{1}{-\Delta x} =$$

$$= (12.63 -_2 12.69) \times_2 \frac{1}{-\Delta x} = -0.06 \times_2 (-50) = 3 = \alpha$$

So,

$$\frac{df}{dx}(4.23)_- = \frac{df}{dx}(4.23)_+ = \frac{df}{dx}(4.23) = 3 = \alpha$$

And we have

$$3 = \alpha +_2 \sigma_4 = 3 +_2 \sigma_4$$

i.e.

$$\sigma_4 = 0$$

Let's take now

$$x = 4.25$$

We get

$$\frac{df}{dx}(4.25)_+ = (3 \times_2 4.27 -_2 3 \times_2 4.25) \times_2 \frac{1}{\Delta x} =$$

$$= (12.81 -_2 12.75) \times_2 \frac{1}{\Delta x} = 0.06 \times_2 50 = 3 = \alpha$$

$$\frac{df}{dx}(4.25)_- = (3 \times_2 4.23 -_2 3 \times_2 4.25) \times_2 \frac{1}{-\Delta x} =$$

$$= (12.69 -_2 12.75) \times_2 \frac{1}{-\Delta x} = -0.06 \times_2 (-50) = 3 = \alpha$$

So,

$$\frac{df}{dx}(4.25)_- = \frac{df}{dx}(4.25)_+ = \frac{df}{dx}(4.25) = 3 = \alpha$$

So, we have again

$$3 = \alpha +_2 \sigma_4 = 3 +_2 \sigma_4$$

i.e.

$$\sigma_4 = 0$$

Example 11:

Let's consider function

$$y = f(x) = x \times_n x; x, y \in W_n$$

Let's take

$$n = 2; \Delta x = 0.02.$$

Direct calculations show

$$\frac{1}{\Delta x} = \frac{1}{0.02} = 50.00, 50.01, \ldots, 50.99$$

We are looking for the right derivatives

$$\frac{df}{dx}(x)_+$$

of function $y = f(x)$ at any point

$$x \in [0.02, 9.97]$$

Direct calculations show that for

$$x \in [0.02, 0.07] \cup [0.10, 0.17] \cup \ldots \cup [0.90, 0.97]$$

$$f(x +_2 \Delta x) -_2 f(x) = 0$$

and

$$\frac{df}{dx}(x)_+ = 0$$

For

$$x = 0.08, 0.09$$

$$f(x +_2 \Delta x) -_2 f(x) = 0.01$$

and

$$\frac{df}{dx}(x)_+ = 0.50$$

For

$$x = 0.18, 0.19$$

$$f(x +_2 \Delta x) -_2 f(x) = 0.03$$

and

$$\frac{df}{dx}(x)_+ = 1.50$$

For

$$x = 0.28, 0.29$$

$$f(x +_2 \Delta x) -_2 f(x) = 0.05$$

and

$$\frac{df}{dx}(x)_+ = 2.50$$

For

$$x = 0.38, 0.39$$

$$f(x +_2 \Delta x) -_2 f(x) = 0.07$$

and

$$\frac{df}{dx}(x)_+ = 3.50$$

For

$$x = 0.48, 0.49$$

$$f(x +_2 \Delta x) -_2 f(x) = 0.09$$

and

$$\frac{df}{dx}(x)_+ = 4.50$$

For

$$x = 0.58, 0.59$$

$$f(x +_2 \Delta x) -_2 f(x) = 0.11$$

and

$$\frac{df}{dx}(x)_+ = 5.50, 5.51, 5.52, 5.53, 5.54, 5.55, 5.56, 5.57, 5.58, 5.59$$

For

$$x = 0.68, 0.69$$

$$f(x +_2 \Delta x) -_2 f(x) = 0.13$$

and

$$\frac{df}{dx}(x)_+ = 6.50, 6.51, 6.52, 6.53, 6.54, 6.55, 6.56, 6.57, 6.58, 6.59$$

For

$$x = 0.78, 0.79$$

$$f(x +_2 \Delta x) -_2 f(x) = 0.15$$

and

$$\frac{df}{dx}(x)_+ = 7.50, 7.51, 7.52, 7.53, 7.54, 7.55, 7.56, 7.57, 7.58, 7.59$$

For

$$x = 0.88, 0.89$$

$$f(x +_2 \Delta x) -_2 f(x) = 0.17$$

and

$$\frac{df}{dx}(x)_+ = 8.50, 8.51, 8.52, 8.53, 8.54, 8.55, 8.56, 8.57, 8.58, 8.59$$

For

$$x = 0.98$$

$$f(x +_2 \Delta x) -_2 f(x) = 0.19$$

and

$$\frac{df}{dx}(x)_+ = 9.50, 9.51, 9.52, 9.53, 9.54, 9.55, 9.56, 9.57, 9.58, 9.59$$

For $x = 0.99$

$$f(x +_2 \Delta x) -_2 f(x) = 0.21$$

and

$$\frac{df}{dx}(x)_+ = 10.50, 10.52, 10.54, 10.56, 10.58, 10.60, 10.62, 10.64, 10.66, 10.68$$

$$x \in [1.00, 1.07] \cup [1.10, 1.17] \cup \ldots \cup [1.90, 1.97]$$

$$f(x +_2 \Delta x) -_2 f(x) = 0.04$$

and

$$\frac{df}{dx}(x)_+ = 2.00$$

For $x = 1.98$

$$f(x +_2 \Delta x) -_2 f(x) = 0.23$$

and

$$\frac{df}{dx}(x)_+ = 11.50, 11.52, 11.54, 11.56, 11.58, 11.60, 11.62, 11.64, 11.66, 11.68$$

For $x = 1.99$

$$f(x +_2 \Delta x) -_2 f(x) = 0.25$$

and

$$\frac{df}{dx}(x)_+ = 12.50, 12.52, 12.54, 12.56, 12.58, 12.60, 12.62, 12.64, 12.66, 12.68$$

Direct calculations show that for

$$x \in [0.02, 2.98]$$

right derivative exists and is uniquely defined in 251 points, and is not uniquely defined (ten different values in each point) in 46 points. For the rest points of segment $[0.02, 9.97]$ right derivative exist and is not uniquely defined (ten different values in each point).

Example 12:

Let's take

$$n = 2; \Delta x = 0.04.$$

Direct calculations show

$$\frac{1}{\Delta x} = \frac{1}{0.04} = 25.00, 25.01, \ldots, 25.99$$

We are looking for the right derivatives

$$\frac{df}{dx}(x)_+$$

of function $y = f(x)$ at any point

$$x \in [0.04, 9.95]$$

Direct calculations show that for

$$x \in [0.04, 0.05] \cup [0.10, 0.15] \cup \ldots \cup [0.90, 0.95]$$

$$f(x +_2 \Delta x) -_2 f(x) = 0$$

and

$$\frac{df}{dx}(x)_+ = 0$$

For

$$x = 0.06, 0.09$$

$$f(x +_2 \Delta x) -_2 f(x) = 0.01$$

and

$$\frac{df}{dx}(x)_+ = 0.25$$

For

$$x = 0.16, 0.19$$

$$f(x +_2 \Delta x) -_2 f(x) = 0.03$$

and

$$\frac{df}{dx}(x)_+ = 0.75$$

For

$$x = 0.26, 0.29$$

$$f(x +_2 \Delta x) -_2 f(x) = 0.05$$

and

$$\frac{df}{dx}(x)_+ = 1.25$$

For

$$x = 0.36, 0.39$$

$$f(x +_2 \Delta x) -_2 f(x) = 0.07$$

and

$$\frac{df}{dx}(x)_+ = 1.75$$

For

$$x = 0.46, 0.49$$

$$f(x +_2 \Delta x) -_2 f(x) = 0.09$$

and

$$\frac{df}{dx}(x)_+ = 2.25$$

For

$$x = 0.56, 0.59$$

$$f(x +_2 \Delta x) -_2 f(x) = 0.11$$

and

$$\frac{df}{dx}(x)_+ = 2.75, 2.76, 2.77, 2.78, 2.79, 2.80, 2.81, 2.82, 2.83, 2.84$$

For

$$x = 0.66, 0.69$$

$$f(x +_2 \Delta x) -_2 f(x) = 0.13$$

and

$$\frac{df}{dx}(x)_+ = 3.25, 3.26, 3.27, 3.28, 3.29, 3.30, 3.31, 3.32, 3.33, 3.34$$

For

$$x = 0.76, 0.79$$

$$f(x +_2 \Delta x) -_2 f(x) = 0.15$$

and

$$\frac{df}{dx}(x)_+ = 3.75, 3.76, 3.77, 3.78, 3.79, 3.80, 3.81, 3.82, 3.83, 3.84$$

For

$$x = 0.86, 0.89$$

$$f(x +_2 \Delta x) -_2 f(x) = 0.17$$

and

$$\frac{df}{dx}(x)_+ = 4.25, 4.26, 4.27, 4.28, 4.29, 4.30, 4.31, 4.32, 4.33, 4.34$$

For

$$x = 0.96$$

$$f(x +_2 \Delta x) -_2 f(x) = 0.19$$

and

$$\frac{df}{dx}(x)_+ = 4.75, 4.76, 4.77, 4.78, 4.79, 4.80, 4.81, 4.82, 4.83, 4.84$$

For $x = 0.97$

$$f(x +_2 \Delta x) -_2 f(x) = 0.21$$

and

$$\frac{df}{dx}(x)_+ = 5.25, 5.27, 5.29, 5.31, 5.33, 5.35, 5.37, 5.39, 5.41, 5.43$$

For $x = 0.98$

$$f(x +_2 \Delta x) -_2 f(x) = 0.23$$

and

$$\frac{df}{dx}(x)_+ = 5.75, 5.77, 5.79, 5.81, 5.83, 5.85, 5.87, 5.89, 5.91, 5.93$$

For $x = 0.99$

$$f(x +_2 \Delta x) -_2 f(x) = 0.25$$

and

$$\frac{df}{dx}(x)_+ = 6.25, 6.27, 6.29, 6.31, 6.33, 6.35, 6.37, 6.39, 6.41, 6.43$$

For

$$x \in [1.00, 1.05] \cup [1.10, 1.15] \cup \ldots \cup [1.90, 1.95]$$

$$f(x +_2 \Delta x) -_2 f(x) = 0.08$$

and

$$\frac{df}{dx}(x)_+ = 2.00$$

For

$$x = 1.96$$

$$f(x +_2 \Delta x) -_2 f(x) = 0.27$$

and

$$\frac{df}{dx}(x)_+ = 6.75, 6.77, 6.79, 6.81, 6.83, 6.85, 6.87, 6.89, 6.91, 6.93$$

For $x = 1.97$

$$f(x +_2 \Delta x) -_2 f(x) = 0.29$$

and

$$\frac{df}{dx}(x)_+ = 7.25, 7.27, 7.29, 7.31, 7.33, 7.35, 7.37, 7.39, 7.41, 7.43$$

For $x = 1.98$

$$f(x +_2 \Delta x) -_2 f(x) = 0.31$$

and

$$\frac{df}{dx}(x)_+ = 7.75, 7.78, 7.81, 7.84, 7.87, 7.90, 7.93, 7.96, 7.99, 8.02$$

For $x = 1.99$

$$f(x +_2 \Delta x) -_2 f(x) = 0.33$$

and

$$\frac{df}{dx}(x)_+ = 8.25, 8.28, 8.31, 8.34, 8.37, 8.40, 8.43, 8.46, 8.49, 8.52$$

Direct calculations show that for

$$x \in [0.04, 1.95]$$

right derivative exist and is uniquely defined in 140 points, and is not uniquely defined (ten different values in each point) in 52 points. For the rest points of segment $[0.04, 9.95]$ right derivative exist and is not uniquely defined (ten different values in each point).

5

Classic Fluid Mechanics Equations and Observability

Classic fluid mechanics was developed from the physical conservation laws of mass, momentum and energy. The control volume is a basic concept in the fluid mechanics, to which the physical conservation laws are applied. In particular the Navier – Stokes equations, developed in 1822, are equations which can be used to determine the velocity vector field that applies to a fluid, given some initial conditions. They arise from the application of Newton's second law in combination with a fluid stress (due to viscosity) and a pressure term. The well-established Taylor series expansion, derivatives and integrals are basic tools used in the classic fluid mechanics.

The set of equations describing classic fluid mechanics includes thermodynamical equations, continuity equation, Euler equation of motion of the fluid, energy flux and moment flux equations, incompressible fluids equations, Navier-Stokes equations.

For example, classic equation of continuity (see [7]) is

$$\partial\rho/\partial t + div(\rho \cdot \mathbf{v}) = 0$$

where $\mathbf{v}(x, y, z, t)$ is the ideal fluid velocity and $p(x, y, z, t), \rho(x, y, z, t)$ are the thermodynamic quantities – pressure and density.

Euler equation (see [7]) is

$$\rho \cdot (\partial\mathbf{v}/\partial t + (\mathbf{v}, \nabla) \times_n \mathbf{v}) = -\mathbf{grad}p$$

where

$$\nabla = (\partial/\partial x, \partial/\partial y, \partial/\partial z)$$

is formal vector with partial derivatives.

Classic fluid mechanics uses thermodynamical parameters and laws (see [10]). Thermodynamics considers thermal natural variables: T – temperature and S – entropy, and mechanical natural variables: P – pressure and V – volume. The combined classic first and second thermodynamic laws give equation

$$\Delta U = T \cdot \Delta S - P \cdot \Delta V$$

Definition of enthalpy can be written as

$$H = U + P \cdot V$$

DOI: 10.1201/9781003175902-5

Definition of Helmholtz free energy can be written as

$$A = U + T \cdot S$$

Definition of Gibbs free energy can be written as

$$G = H + T \cdot S$$

Energy flux equation (see [7]) is

$$\partial/\partial t(\frac{1}{2} \cdot (\rho \cdot (\mathbf{v}, \mathbf{v})) + \rho \cdot \epsilon) =$$

$$= -(\frac{1}{2} \cdot (\mathbf{v}, \mathbf{v})) \cdot div(\rho \cdot \mathbf{v}) - (\mathbf{v}, \mathbf{grad}\ p) - (\rho \cdot \mathbf{v}, (\mathbf{v}, \nabla) \cdot \mathbf{v}) -$$

$$-div(\rho \cdot \mathbf{v}) \cdot \epsilon + \rho \cdot \partial \epsilon/\partial t$$

where ϵ is internal energy per unit mass.

Navier-Stokes equations for incompressible fluid in gravitation field (see [7]) are represented by the following system:

$$\begin{cases} \rho \cdot (\partial \mathbf{v}/\partial t + (\mathbf{v}, \nabla) \cdot \mathbf{v}) = -\mathbf{grad}p + \rho \cdot \mathbf{g} + \alpha \cdot \Delta\mathbf{v} \\ div\mathbf{v} = 0 \end{cases}$$

where

$$\Delta = \partial^2/\partial x_1^2 +_n \partial^2/\partial x_2^2 +_n \partial^2/\partial x_3^2$$

is a Laplace transformation.

Going to the Observer's point of view let's consider first the main Observer's tools which are necessary for the remodeling of the fluid mechanics laws.

We consider below derivatives (and partial derivatives) in three different meanings – right derivatives, left derivatives and both-sides derivatives (or derivatives). So, each statement below has three different meanings.

(D1)

Let f, f_1 and f_2 be the differentiable functions of several variables $x, y, z, ...$; and $f, f_1, f_2, x, y, z, .. \in W_n$; and all partial derivatives belong to W_n. We have

$$\partial f/\partial x = (f(x +_n \Delta x, y, z, ...) - f(x, y, z, ...)) \times_n (\frac{1}{\Delta x})$$

So, we have

$$((f_1 +_n f_2)(x +_n \Delta x, y, z, ...) -_n (f_1 +_n f_2)(x, y, z, ...)) \times_n (\frac{1}{\Delta x}) =$$

$$= ((f_1(x +_n \Delta x, y, z, ...) +_n f_2(x +_n \Delta x, y, z, ...)) -_n$$

$$-_n(f_1(x, y, z, ...) +_n f_2(x, y, z, ...))) \times_n (\frac{1}{\Delta x}) =$$

$$= ((f_1(x +_n \Delta x, y, z, ...) -_n f_1(x, y, z, ...)) +_n$$

$$+_n (f_2(x +_n \Delta x, y, z, ...) -_n f_2(x, y, z, ...))) \times_n (\frac{1}{\Delta x}) =$$

$$= (f_1(x +_n \Delta x, y, z, ...) -_n f_1(x, y, z, ...)) \times_n (\frac{1}{\Delta x}) +_n$$

$$+_n (f_2(x +_n \Delta x, y, z, ...) -_n f_2(x, y, z, ...)) \times_n (\frac{1}{\Delta x}) +_n \xi_1 =$$

$$= \partial f_1/\partial x +_n \partial f_2/\partial x +_n \xi_1$$

And finally

$$\partial(f_1 +_n f_2)/\partial x = \partial f_1/\partial x +_n \partial f_2/\partial x +_n \xi_1.$$

We have to assume that all elements of these equalities are in W_n. Here ξ_1 is a random variable depends on n and m. This variable appears because we know that

$$\delta_2 = (\alpha +_n \beta) \times_n \gamma -_n (\alpha \times_n \gamma +_n \beta \times_n \gamma),$$

$$\alpha, \beta, \gamma \in W_n$$

is a random variable in W_n, and $\delta_2 = 0$ with probability $P < 1$. Also note that a number $\frac{1}{\Delta x}$ exists with probability less than 1. So, probability of equality

$$\partial(f_1 +_n f_2)/\partial x = \partial f_1/\partial x +_n \partial f_2/\partial x$$

is less than 1.

So, we proved

Theorem 5.1 *Let f, f_1 and f_2 be the differentiable functions of several variables $x, y, z, ...$; and $f, f_1, f_2, x, y, z, .. \in W_n$; and all partial derivatives belong to W_n. Then*

$$\partial(f_1 +_n f_2)/\partial x = \partial f_1/\partial x +_n \partial f_2/\partial x +_n \xi_1.$$

Corollary 5.2 *The probability of equality*

$$\partial(f_1 +_n f_2)/\partial x = \partial f_1/\partial x +_n \partial f_2/\partial x$$

is less than 1.

(D2)

Let f_1 and f_2 be the differentiable functions of several variables $x, y, z, ...$; and $f_1, f_2, x, y, z, .. \in W_n$; and all partial derivatives belong to W_n. We have

$$((f_1 \times_n f_2)(x +_n \Delta x, y, z, ...) -_n (f_1 \times_n f_2)(x, y, z, ...)) \times_n (\frac{1}{\Delta x}) =$$

$$= (f_1(x +_n \Delta x, y, z, ...) \times_n f_2(x +_n \Delta x, y, z, ...) -_n$$

$$-_n f_1(x, y, z, ...) \times_n f_2(x, y, z, ...)) \times_n (\frac{1}{\Delta x}) =$$

$$= (f_1(x +_n \Delta x, y, z, ...) \times_n f_2(x +_n \Delta x, y, z, ...) -_n$$

$$-_n f_1(x, y, z, ...) \times_n f_2(x +_n \Delta x, y, z, ...) +_n$$

$$+_n f_1(x, y, z, ...) \times_n f_2(x +_n \Delta x, y, z, ...) -_n$$

$$-_n f_1(x, y, z, ...) \times_n f_2(x, y, z, ...)) \times_n (\frac{1}{\Delta x}) =$$

$$= ((f_1(x +_n \Delta x, y, z, ...) -_n f_1(x, y, z, ...)) \times_n f_2(x +_n \Delta x, y, z, ...) +_n \xi_8 +_n$$

$$+_n (f_2(x +_n \Delta x, y, z, ...) -_n f_2(x, y, z, ...)) \times_n f_1(x, y, z, ...) +_n \xi_9) \times_n (\frac{1}{\Delta x}) =$$

$$= (f_1(x +_n \Delta x, y, z, ...) -_n f_1(x, y, z, ...)) \times_n (\frac{1}{\Delta x}) \times_n f_2(x +_n \Delta x, y, z, ...) +_n$$

$$+_n \xi_8 \times_n (\frac{1}{\Delta x}) +_n$$

$$+_n (f_2(x +_n \Delta x, y, z, ...) -_n f_2(x, y, z, ...)) \times_n (\frac{1}{\Delta x}) \times_n f_1(x, y, z, ...) +_n$$

$$+_n \xi_9 \times_n (\frac{1}{\Delta x}) +_n \xi_{10} =$$

$$= \partial f_1/\partial x \times_n f_2(x +_n \Delta x, y, z, ...) +_n \partial f_2/\partial x \times_n f_1(x, y, z, ...) +_n$$

$$+_n \xi_8 \times_n (\frac{1}{\Delta x}) +_n \xi_9 \times_n (\frac{1}{\Delta x}) +_n \xi_{10} =$$

$$= \partial f_1/\partial x \times_n f_2(x, y, z, ...) +_n \partial f_2/\partial x \times_n f_1(x, y, z, ...) +_n$$

$$+_n \xi_8 \times_n (\frac{1}{\Delta x}) +_n \xi_9 \times_n (\frac{1}{\Delta x}) +_n \xi_{10} +_n$$

$$+_n \partial f_1/\partial x \times_n \partial f_2/\partial x \times_n \Delta x +_n \xi_{11}^1$$

So, we have

$$\partial(f_1 \times_n f_2)/\partial x = \partial f_1/\partial x \times_n f_2 +_n \partial f_2/\partial x \times_n f_1 +_n \xi_{11}^2$$

where

$$\xi_{11}^2 = \xi_8 \times_n (\frac{1}{\Delta x}) +_n \xi_9 \times_n (\frac{1}{\Delta x}) +_n \xi_{10} +_n \partial f_1/\partial x \times_n \partial f_2/\partial x \times_n \Delta x +_n \xi_{11}^1$$

and $\xi_8, \xi_9, \xi_{10}, \xi_{11}^1$ are the random variables depend on n and m, because we know that

$$\delta_2 = (\alpha +_n \beta) \times_n \gamma -_n (\alpha \times_n \gamma +_n \beta \times_n \gamma), \alpha, \beta, \gamma \in W_n$$

$$\delta_3 = \alpha \times_n (\beta \times_n \gamma) -_n (\alpha \times_n \beta) \times_n \gamma, (\alpha, \beta, \gamma \in W_n)$$

are the random variables in W_n, and $\delta_2 = \delta_3 = 0$ with probability $P < 1$. Also note that a number $\frac{1}{\Delta x}$ exists with probability less than 1. So, probability of equality

$$\partial(f_1 \times_n f_2)/\partial x = \partial f_1/\partial x \times_n f_2 +_n \partial f_2/\partial x \times_n f_1$$

is less than 1.

So, we proved

Theorem 5.3 *Let f_1 and f_2 be the differentiable functions of several variables $x, y, z, ...;$ and $f_1, f_2, x, y, z, .. \in W_n$; and all partial derivatives belong to W_n. Then*

$$\partial(f_1 \times_n f_2)/\partial x = \partial f_1/\partial x \times_n f_2 +_n \partial f_2/\partial x \times_n f_1 +_n \xi_{11}^2$$

Corollary 5.4 *The probability of equality*

$$\partial(f_1 \times_n f_2)/\partial x = \partial f_1/\partial x \times_n f_2 +_n \partial f_2/\partial x \times_n f_1$$

is less than 1.

(D3)

Let f be the differentiable function of several variables $x, y, z, ...$ and $\alpha = const$; and $f, x, y, z, .., \alpha \in W_n$; and all partial derivatives belong to W_n. We can calculate and compare two expressions:

$$\partial(\alpha \times_n f)/\partial x$$

and

$$\alpha \times_n \partial f/\partial x$$

We have

$$(\alpha \times_n f(x +_n \Delta x, y, z, ...) -_n \alpha \times_n f(x, y, z, ...)) \times_n (\frac{1}{\Delta x}) =$$

$$= (\alpha \times_n (f(x +_n \Delta x, y, z, ...) -_n f(x, y, z, ...)) +_n \xi_{12}) \times_n (\frac{1}{\Delta x}) =$$

$$= (\alpha \times_n (f(x +_n \Delta x, y, z, ...) -_n f(x, y, z, ...)) \times_n (\frac{1}{\Delta x}) +_n \xi_{12} \times_n (\frac{1}{\Delta x}) +_n \xi_{13}$$

So, we have

$$\partial(\alpha \times_n f)/\partial x = \alpha \times_n \partial f/\partial x + \xi_{14}$$

where ξ_{14} is a random variable depends on n and m, because we know that

$$\delta_2 = (\alpha +_n \beta) \times_n \gamma -_n (\alpha \times_n \gamma +_n \beta \times_n \gamma), \alpha, \beta, \gamma \in W_n$$

$$\delta_3 = \alpha \times_n (\beta \times_n \gamma) -_n (\alpha \times_n \beta) \times_n \gamma, (\alpha, \beta, \gamma \in W_n)$$

are the random variables in W_n, and $\delta_2 = \delta_3 = 0$ with probability $P < 1$. And

$$\xi_{14} = \xi_{12} \times_n (\frac{1}{\Delta x}) +_n \xi_{13}$$

So, probability of equality

$$\partial(\alpha \times_n f)/\partial x = \alpha \times_n \partial f/\partial x$$

is less than 1.

So, we proved

Theorem 5.5 *Let f be the differentiable function of several variables $x, y, z, \ldots$ and $\alpha = const$; and $f, x, y, z, \ldots, \alpha \in W_n$; and all partial derivatives belong to W_n. Then*

$$\partial(\alpha \times_n f)/\partial x = \alpha \times_n \partial f/\partial x + \xi_{14}$$

Corollary 5.6 *The probability of equality*

$$\partial(\alpha \times_n f)/\partial x = \alpha \times_n \partial f/\partial x$$

is less than 1.

(D4)

Let f be the differentiable function of several variables $x, y, z, \ldots$, and $f, x, y, z, \ldots \in W_n$, and all partial derivatives belong to W_n. Let's consider and compare $\partial/\partial x(\partial f/\partial y)$ and $\partial/\partial y(\partial f/\partial x)$ Then

$$\partial f/\partial y = (f(x, y +_n \Delta y, z, \ldots) -_n f(x, y, z, \ldots)) \times_n \left(\frac{1}{\Delta y}\right)$$

$$\partial/\partial x(\partial f/\partial y) = ((f(x+_n\Delta x, y+_n\Delta y, z, \ldots) -_n f(x+_n\Delta x, y, z, \ldots)) \times_n \left(\frac{1}{\Delta y}\right) -_n$$

$$-_n(f(x, y +_n \Delta y, z, \ldots) -_n f(x, y, z, \ldots)) \times_n \left(\frac{1}{\Delta y}\right)) \times_n \left(\frac{1}{\Delta x}\right) =$$

$$= ((f(x+_n\Delta x, y+_n\Delta y, z, \ldots) -_n f(x, y+_n\Delta y, z, \ldots)) -_n (f(x+_n\Delta x, y, z, \ldots) -_n$$

$$-_n f(x, y, z, \ldots)) \times_n \left(\frac{1}{\Delta y}\right) +_n \xi_2) \times_n \left(\frac{1}{\Delta x}\right) =$$

$$= ((f(x+_n\Delta x, y+_n\Delta y, z, \ldots) -_n f(x, y+_n\Delta y, z, \ldots)) -_n (f(x+_n\Delta x, y, z, \ldots) -_n$$

$$-_n f(x, y, z, \ldots)) \times_n \left(\frac{1}{\Delta y}\right)) \times_n \left(\frac{1}{\Delta x}\right) +_n$$

$$+_n \xi_2 \times_n \left(\frac{1}{\Delta x}\right) +_n \xi_3 =$$

$$= ((f(x +_n \Delta x, y +_n \Delta y, z, \ldots) -_n f(x, y +_n \Delta y, z, \ldots)) \times_n \left(\frac{1}{\Delta x}\right) -_n$$

$$-_n(f(x +_n \Delta x, y, z, \ldots) -_n f(x, y, z, \ldots)) \times_n \left(\frac{1}{\Delta x}\right) +_n$$

$$+_n \xi_4) \times_n \left(\frac{1}{\Delta y}\right) +_n \xi_2 \times_n \left(\frac{1}{\Delta x}\right) +_n \xi_3 =$$

$$= ((f(x +_n \Delta x, y +_n \Delta y, z, \ldots) -_n f(x, y +_n \Delta y, z, \ldots)) \times_n \left(\frac{1}{\Delta x}\right) -_n$$

$$-_n(f(x +_n \Delta x, y, z, \ldots) -_n f(x, y, z, \ldots)) \times_n \left(\frac{1}{\Delta x}\right)) \times_n \left(\frac{1}{\Delta y}\right) +_n \xi_4 \times_n \left(\frac{1}{\Delta y}\right) +_n$$

$$+_n \xi_5 +_n \xi_2 \times_n \left(\frac{1}{\Delta x}\right) +_n \xi_3$$

And

$$\partial f / \partial x = (f(x +_n \Delta x, y, z, ...) - f(x, y, z, ...)) \times_n \left(\frac{1}{\Delta x}\right)$$

$$\partial / \partial y (\partial f / \partial x) = ((f(x +_n \Delta x, y +_n \Delta y, z, ...) - f(x, y +_n \Delta y, z, ...)) \times_n \left(\frac{1}{\Delta x}\right) -_n$$

$$-_n (f(x +_n \Delta x, y, z, ...) - f(x, y, z, ...)) \times_n \left(\frac{1}{\Delta x}\right)) \times_n \left(\frac{1}{\Delta y}\right)$$

We have to assume that all elements of these equalities are in W_n. That means

$$\partial / \partial x (\partial f / \partial y) = \partial / \partial y (\partial f / \partial x) +_n \xi_6$$

and ξ_6 is a random variable depends on n and m. Let's note that random variables $\xi_2, \xi_3, \xi_4, \xi_5$ appear because we know that

$$\delta_2 = (\alpha +_n \beta) \times_n \gamma -_n (\alpha \times_n \gamma +_n \beta \times_n \gamma), \alpha, \beta, \gamma \in W_n$$

$$\delta_3 = \alpha \times_n (\beta \times_n \gamma) -_n (\alpha \times_n \beta) \times_n \gamma, (\alpha, \beta, \gamma \in W_n)$$

are the random variables in W_n, and $\delta_2 = \delta_3 = 0$ with probability $P < 1$. Also note that numbers $\frac{1}{\Delta x}, \frac{1}{\Delta y}$ exist with probability less than 1. So, probability of equality

$$\partial / \partial x (\partial f / \partial y) = \partial / \partial y (\partial f / \partial x)$$

is less than 1.

So, we proved

Theorem 5.7 *Let f be the differentiable function of several variables $x, y, z, ...,$ and $f, x, y, z, .. \in W_n$, and all partial derivatives belong to W_n Then*

$$\partial / \partial x (\partial f / \partial y) = \partial / \partial y (\partial f / \partial x) +_n \xi_6$$

Corollary 5.8 *The probability of equality*

$$\partial / \partial x (\partial f / \partial y) = \partial / \partial y (\partial f / \partial x)$$

is less than 1.

(D5)
We have vector field

$$\mathbf{F} = (f_1(x, y, z), f_2(x, y, z), f_3(x, y, z))$$

with f_1, f_2, f_3, x, y, z belong to W_n. We assume that f_1, f_2, f_3 are differentiable functions, and all partial derivatives belong to W_n.

And we can consider

$$\mathbf{rot}\mathbf{F} = \nabla \times \mathbf{F} = (\partial f_3/\partial y -_n \partial f_2/\partial z) \times_n \mathbf{i} -_n (\partial f_3/\partial x -_n \partial f_1/\partial z) \times_n \mathbf{j} +_n$$

$$+_n(\partial f_2/\partial x -_n \partial f_1/\partial y) \times_n \mathbf{k}$$

If we have another vector field

$$\mathbf{G} = (g_1(x,y,z), g_2(x,y,z), g_3(x,y,z))$$

we can calculate and compare

$$\mathbf{rot}\mathbf{F}, \mathbf{rot}\mathbf{G}, \mathbf{rot}(\mathbf{F} +_n \mathbf{G})$$

We have

$$\mathbf{rot}(\mathbf{F} +_n \mathbf{G}) = (\partial(f_3 +_n g_3)/\partial y -_n \partial(f_2 +_n g_2)/\partial z) \times_n \mathbf{i} -_n (\partial(f_3 +_n g_3)/\partial x -_n$$

$$-_n \partial(f_1 +_n g_1)/\partial z) \times_n \mathbf{j} +_n (\partial(f_2 +_n g_2)/\partial x -_n \partial(f_1 +_n g_1)/\partial y) \times_n \mathbf{k}$$

By **(D1)** we get

$$\mathbf{rot}(\mathbf{F} +_n \mathbf{G}) = \mathbf{rot}\mathbf{F} +_n \mathbf{rot}\mathbf{G} +_n \xi_7$$

where ξ_7 is a random vector depends on n and m, because we know that

$$\delta_2 = (\alpha +_n \beta) \times_n \gamma -_n (\alpha \times_n \gamma +_n \beta \times_n \gamma), \alpha, \beta, \gamma \in W_n$$

$$\delta_3 = \alpha \times_n (\beta \times_n \gamma) -_n (\alpha \times_n \beta) \times_n \gamma, (\alpha, \beta, \gamma \in W_n)$$

are the random variables in W_n, and $\delta_2 = \delta_3 = 0$ with probability $P < 1$. So, probability of equality

$$\mathbf{rot}(\mathbf{F} +_n \mathbf{G}) = \mathbf{rot}\mathbf{F} +_n \mathbf{rot}\mathbf{G}$$

is less than 1.
So, we proved

Theorem 5.9 *We have vector fields*

$$\boldsymbol{F} = (f_1(x,y,z), f_2(x,y,z), f_3(x,y,z))$$

and

$$\boldsymbol{G} = (g_1(x,y,z), g_2(x,y,z), g_3(x,y,z))$$

with $f_1, f_2, f_3, g_1, g_2, g_3, x, y, z$ belong to W_n. We assume that $f_1, f_2, f_3, g_1, g_2, g_3$ are differentiable functions, and all partial derivatives belong to W_n. Then

$$\boldsymbol{rot}(\boldsymbol{F} +_n \boldsymbol{G}) = \boldsymbol{rot}\boldsymbol{F} +_n \boldsymbol{rot}\boldsymbol{G} +_n \xi_7$$

Corollary 5.10 *The probability of equality*

$$rot(\mathbf{F} +_n \mathbf{G}) = rot\mathbf{F} +_n rot\mathbf{G}$$

is less than 1.

(D6)

We have vector field

$$\mathbf{F} = (f_1(x, y, z), f_2(x, y, z), f_3(x, y, z))$$

with f_1, f_2, f_3, x, y, z belong to W_n. We assume that f_1, f_2, f_3 are differentiable functions, and all partial derivatives belong to W_n.

And we can consider

$$\mathbf{rotF} = \nabla \times \mathbf{F} = (\partial f_3/\partial y -_n \partial f_2/\partial z) \times_n \mathbf{i} -_n (\partial f_3/\partial x -_n \partial f_1/\partial z) \times_n \mathbf{j} +_n$$

$$+_n(\partial f_2/\partial x -_n \partial f_1/\partial y) \times_n \mathbf{k}$$

Let's take $\alpha = const \in W_n$ and consider vector field

$$\alpha \times_n \mathbf{F} = (\alpha \times_n f_1(x, y, z), \alpha \times_n f_2(x, y, z), \alpha \times_n f_3(x, y, z))$$

We have

$$\mathbf{rot}(\alpha \times_n \mathbf{F}) =$$

$$= (\partial(\alpha \times_n f_3)/\partial y -_n \partial(\alpha \times_n f_2)/\partial z) \times_n \mathbf{i} -_n$$

$$-_n(\partial(\alpha \times_n f_3)/\partial x -_n \partial(\alpha \times_n f_1)/\partial z) \times_n \mathbf{j} +_n$$

$$+_n(\partial(\alpha \times_n f_2)/\partial x -_n \partial(\alpha \times_n f_1)/\partial y) \times_n \mathbf{k}$$

By **(D3)** we get

$$\mathbf{rot}(\alpha \times_n \mathbf{F}) = \alpha \times_n \mathbf{rotF} +_n \xi_{15}$$

where ξ_{15} is a random vector depends on n and m, because we know that

$$\delta_2 = (\alpha +_n \beta) \times_n \gamma -_n (\alpha \times_n \gamma +_n \beta \times_n \gamma), \alpha, \beta, \gamma \in W_n$$

$$\delta_3 = \alpha \times_n (\beta \times_n \gamma) -_n (\alpha \times_n \beta) \times_n \gamma, (\alpha, \beta, \gamma \in W_n)$$

are the random variables in W_n, and $\delta_2 = \delta_3 = 0$ with probability $P < 1$.

So, probability of equality

$$\mathbf{rot}(\alpha \times_n \mathbf{F}) = \alpha \times_n \mathbf{rotF}$$

is less than 1.

So, we proved

Theorem 5.11 *We have vector field*

$$\boldsymbol{F} = (f_1(x,y,z), f_2(x,y,z), f_3(x,y,z))$$

with f_1, f_2, f_3, x, y, z belong to W_n. We assume that f_1, f_2, f_3 are differentiable functions, and all partial derivatives belong to W_n. And we have $\alpha = const \in W_n$. Then

$$\boldsymbol{rot}(\alpha \times_n \boldsymbol{F}) = \alpha \times_n \boldsymbol{rotF} +_n \xi_{15}$$

Corollary 5.12 *The probability of equality*

$$\boldsymbol{rot}(\alpha \times_n \boldsymbol{F}) = \alpha \times_n \boldsymbol{rotF}$$

is less than 1.

(D7)

Let f be the differentiable function of variables x, y, z; and $f, x, y, z \in W_n$; and all partial derivatives belong to W_n.

We call "gradient of function f" the following vector

$$\mathbf{grad}f = \nabla f = (\partial f/\partial x, \partial f/\partial y, \partial f/\partial z)$$

For vector fields $\mathbf{a}, \mathbf{b} \in W_n \times W_n \times W_n$ let's consider and compare two vectors:

$$\mathbf{grad}(\mathbf{a}, \mathbf{b})$$

and

$$(\mathbf{a}, \nabla)\mathbf{b} +_n (\mathbf{b}, \nabla)\mathbf{a} +_n \mathbf{b} \times \mathbf{rota} +_n \mathbf{a} \times \mathbf{rotb}$$

Let's

$$\mathbf{a} = (a_1(x,y,z), a_2(x,y,z), a_3(x,y,z))$$

$$\mathbf{b} = (b_1(x,y,z), b_2(x,y,z), b_3(x,y,z))$$

Then by **(D1)** and **(D2)** we get

$$\mathbf{grad}(\mathbf{a}, \mathbf{b}) =$$

$$= (\partial(a_1 \times_n b_1 +_n a_2 \times_n b_2 +_n a_3 \times_n b_3)/\partial x,$$

$$\partial(a_1 \times_n b_1 +_n a_2 \times_n b_2 +_n a_3 \times_n b_3)/\partial y,$$

$$\partial(a_1 \times_n b_1 +_n a_2 \times_n b_2 +_n a_3 \times_n b_3)/\partial z) =$$

$$= (\partial a_1/\partial x \times_n b_1 +_n a_1 \times_n \partial b_1/\partial x +_n \partial a_2/\partial x \times_n b_2 +_n a_2 \times_n \partial b_2/\partial x +_n$$

$$+_n \partial a_3/\partial x \times_n b_3 +_n a_3 \times_n \partial b_3/\partial x,$$

$$\partial a_1/\partial y \times_n b_1 +_n a_1 \times_n \partial b_1/\partial y +_n \partial a_2/\partial y \times_n b_2 +_n a_2 \times_n \partial b_2/\partial y +_n$$

$$+_n \partial a_3/\partial y \times_n b_3 +_n a_3 \times_n \partial b_3/\partial y,$$

$$\partial a_1/\partial z \times_n b_1 +_n a_1 \times_n \partial b_1/\partial z +_n \partial a_2/\partial z \times_n b_2 +_n a_2 \times_n \partial b_2/\partial z +_n$$

$$+_n \partial a_3/\partial z \times_n b_3 +_n a_3 \times_n \partial b_3/\partial z) +_n$$

$$+_n \xi_{16}$$

where ξ_{16} is a random vector depends on n and m.
Let's consider

$$(\mathbf{a}, \nabla)\mathbf{b} = (a_1 \times_n \partial b_1/\partial x +_n a_2 \times_n \partial b_1/\partial y +_n a_3 \times_n \partial b_1/\partial z,$$

$$a_1 \times_n \partial b_2/\partial x +_n a_2 \times_n \partial b_2/\partial y +_n a_3 \times_n \partial b_2/\partial z,$$

$$a_1 \times_n \partial b_3/\partial x +_n a_2 \times_n \partial b_3/\partial y +_n a_3 \times_n \partial b_3/\partial z)$$

Now let's consider

$$(\mathbf{b}, \nabla)\mathbf{a} = (b_1 \times_n \partial a_1/\partial x +_n b_2 \times_n \partial a_1/\partial y +_n b_3 \times_n \partial a_1/\partial z,$$

$$b_1 \times_n \partial a_2/\partial x +_n b_2 \times_n \partial a_2/\partial y +_n b_3 \times_n \partial a_2/\partial z,$$

$$b_1 \times_n \partial a_3/\partial x +_n b_2 \times_n \partial a_3/\partial y +_n b_3 \times_n \partial a_3/\partial z)$$

Also by **(D1)** and **(D2)** we get

$$\mathbf{b} \times \mathbf{rot a} = \mathbf{b} \times (\nabla \times \mathbf{a}) =$$

$$= \mathbf{b} \times ((\partial a_3/\partial y -_n \partial a_2/\partial z) \times_n \mathbf{i} -_n (\partial a_3/\partial x -_n \partial a_1/\partial z) \times_n \mathbf{j} +_n$$

$$+_n(\partial a_2/\partial x -_n \partial a_1/\partial y) \times_n \mathbf{k}) =$$

$$= (b_2 \times_n (\partial a_2/\partial x -_n \partial a_1/\partial y) -_n b_3 \times_n (-\partial a_3/\partial x +_n \partial a_1/\partial z)) \times_n \mathbf{i} -_n$$

$$-_n(b_1 \times_n (\partial a_2/\partial x -_n \partial a_1/\partial y) -_n b_3 \times_n (\partial a_3/\partial y -_n \partial a_2/\partial z)) \times_n \mathbf{j} +_n$$

$$+_n(b_1 \times_n (-\partial a_3/\partial x +_n \partial a_1/\partial z) -_n b_2 \times_n (\partial a_3/\partial y -_n \partial a_2/\partial z)) \times_n \mathbf{k} =$$

$$= (b_2 \times_n \partial a_2/\partial x -_n b_2 \times_n \partial a_1/\partial y +_n b_3 \times_n \partial a_3/\partial x -_n$$

$$-_n b_3 \times_n \partial a_1/\partial z +_n \xi_{17}) \times_n \mathbf{i} -_n$$

$$-_n(b_1 \times_n \partial a_2/\partial x -_n b_1 \times_n \partial a_1/\partial y -_n b_3 \times_n \partial a_3/\partial y +_n$$

$$+_n b_3 \times_n \partial a_2/\partial z +_n \xi_{18}) \times_n \mathbf{j} +_n$$

$$+_n(-b_1 \times_n \partial a_3/\partial x +_n b_1 \times_n \partial a_1/\partial z -_n b_2 \times_n \partial a_3/\partial y +_n$$

$$+_n b_2 \times_n \partial a_2/\partial z +_n \xi_{19}) \times_n \mathbf{k}$$

where
$\xi_{17}, \xi_{18}, \xi_{19}$ are the random variables depend on n and m,

$$\xi_{23}^1 = \xi_{17} \times_n \mathbf{i} -_n \xi_{18} \times_n \mathbf{j} +_n \xi_{19} \times_n \mathbf{k}$$

is a random vector depends on n and m.
And finally let's consider (also by **(D1)** and **(D2)**)

$$\mathbf{a} \times \mathbf{rotb} = \mathbf{a} \times (\nabla \times \mathbf{b}) =$$

$$= \mathbf{a} \times ((\partial b_3/\partial y -_n \partial b_2/\partial z) \times_n \mathbf{i} -_n (\partial b_3/\partial x -_n \partial b_1/\partial z) \times_n \mathbf{j} +_n$$

$$+_n(\partial b_2/\partial x -_n \partial b_1/\partial y) \times_n \mathbf{k}) =$$

$$= (a_2 \times_n (\partial b_2/\partial x -_n \partial b_1/\partial y) -_n a_3 \times_n (-\partial b_3/\partial x +_n \partial b_1/\partial z)) \times_n \mathbf{i} -_n$$

$$-_n(a_1 \times_n (\partial b_2/\partial x -_n \partial b_1/\partial y) -_n a_3 \times_n (\partial b_3/\partial y -_n \partial b_2/\partial z)) \times_n \mathbf{j} +_n$$

$$+_n(a_1 \times_n (-\partial b_3/\partial x +_n \partial b_1/\partial z) -_n a_2 \times_n (\partial b_3/\partial y -_n \partial b_2/\partial z)) \times_n \mathbf{k} =$$

$$= (a_2 \times_n \partial b_2/\partial x -_n a_2 \times_n \partial b_1/\partial y +_n a_3 \times_n \partial b_3/\partial x -_n$$

$$-_n a_3 \times_n \partial b_1/\partial z +_n \xi_{20}) \times_n \mathbf{i} -_n$$

$$-_n(a_1 \times_n \partial b_2/\partial x -_n a_1 \times_n \partial b_1/\partial y -_n a_3 \times_n \partial b_3/\partial y +_n$$

$$+_n a_3 \times_n \partial b_2/\partial z +_n \xi_{21}) \times_n \mathbf{j} +_n$$

$$+_n(-a_1 \times_n \partial b_3/\partial x +_n a_1 \times_n \partial b_1/\partial z -_n a_2 \times_n \partial b_3/\partial y +_n$$

$$+_n a_2 \times_n \partial b_2/\partial z +_n \xi_{22}) \times_n \mathbf{k}$$

where
$\xi_{20}, \xi_{21}, \xi_{22}$ are the random variables depend on n and m,

$$\xi_{23}^2 = \xi_{20} \times_n \mathbf{i} -_n \xi_{21} \times_n \mathbf{j} +_n \xi_{22} \times_n \mathbf{k}$$

is a random vector depend on n and m.
So, we have

$$\mathbf{grad(a, b)} = (\mathbf{a}, \nabla)\mathbf{b} +_n (\mathbf{b}, \nabla)\mathbf{a} +_n \mathbf{b} \times \mathbf{rota} +_n \mathbf{a} \times \mathbf{rotb} +_n$$

$$+_n \xi_{16} -_n \xi_{23}^1 -_n \xi_{23}^2$$

If we take vector

$$\xi_{24} = \xi_{16} -_n \xi_{23}^1 -_n \xi_{23}^2$$

we can rewrite previous equation as

$$\mathbf{grad(a, b)} = (\mathbf{a}, \nabla)\mathbf{b} +_n (\mathbf{b}, \nabla)\mathbf{a} +_n \mathbf{b} \times \mathbf{rota} +_n \mathbf{a} \times \mathbf{rotb} +_n \xi_{24}$$

So, probability of equality

$$\mathbf{grad(a, b)} = (\mathbf{a}, \nabla)\mathbf{b} +_n (\mathbf{b}, \nabla)\mathbf{a} +_n \mathbf{b} \times \mathbf{rota} +_n \mathbf{a} \times \mathbf{rotb}$$

is less than 1.
So, we proved

Theorem 5.13 *Let f be the differentiable function of variables x, y, z; and $f, x, y, z \in W_n$; and all partial derivatives belong to W_n. And let's consider vector fields $a, b \in W_n \times W_n \times W_n$ Then*

$$grad(a, b) = (a, \nabla)b +_n (b, \nabla)a +_n b \times rota +_n a \times rotb +_n \xi_{24}$$

Corollary 5.14 *The probability of equality*

$$grad(a, b) = (a, \nabla)b +_n (b, \nabla)a +_n b \times rota +_n a \times rotb$$

is less than 1.

(D8)

Let f be the differentiable function of variables x, y, z; and $f, x, y, z \in W_n$; and all partial derivatives belong to W_n.

Gradient of function f is the following vector

$$\mathbf{grad} f = \nabla f = (\partial f / \partial x, \partial f / \partial y, \partial f / \partial z)$$

Let's calculate

$$\mathbf{rotgrad} f = (\partial / \partial y(\partial f / \partial z) -_n \partial / \partial z(\partial f / \partial y)) \times_n \mathbf{i} -_n (\partial / \partial x(\partial f / \partial z) -_n$$

$$-_n \partial / \partial z(\partial f / \partial x)) \times_n \mathbf{j} +_n (\partial / \partial x(\partial f / \partial y) -_n \partial / \partial y(\partial f / \partial x)) \times_n \mathbf{k}$$

Following by **(D4)** we get

$$\mathbf{rotgrad} f = \xi_{25} \times_n \mathbf{i} -_n \xi_{26} \times_n \mathbf{j} +_n \xi_{27} \times_n \mathbf{k}$$

where $\xi_{25}, \xi_{26}, \xi_{27}$ are the random variables depend on n and m. If we introduce random vector

$$\xi_{28} = \xi_{25} \times_n \mathbf{i} -_n \xi_{26} \times_n \mathbf{j} +_n \xi_{27} \times_n \mathbf{k}$$

we can rewrite

$$\mathbf{rotgrad} f = \xi_{28}$$

So, probability of equality

$$\mathbf{rotgrad} f = 0$$

is less than 1.
So, we proved

Theorem 5.15 *Let f be the differentiable function of variables x, y, z; and $f, x, y, z \in W_n$; and all partial derivatives belong to W_n. Then*

$$\mathbf{rotgrad} f = \xi_{28}$$

Corollary 5.16 *The probability of equality*

$$rotgradf = 0$$

is less than 1.

(D9)
We have vector field

$$\mathbf{F} = (f_1(x,y,z), f_2(x,y,z), f_3(x,y,z))$$

with f_1, f_2, f_3, x, y, z belong to W_n. We assume that f_1, f_2, f_3 are differentiable functions, and all partial derivatives belong to W_n.

We consider now divergence - scalar function on $\mathbf{F}$

$$div\mathbf{F} = (\nabla, \mathbf{F}) = \partial f_1/\partial x +_n \partial f_2/\partial y +_n \partial f_3/\partial z$$

Let's consider another vector field

$$\mathbf{G} = (g_1(x,y,z), g_2(x,y,z), g_3(x,y,z))$$

also with g_1, g_2, g_3, x, y, z belong to W_n. We assume that g_1, g_2, g_3 are differentiable functions, and all partial derivatives belong to W_n.

Let's calculate now

$$div(\mathbf{F} +_n \mathbf{G}) = \partial(f_1 +_n g_1)/\partial x +_n \partial(f_2 +_n g_2)/\partial y +_n \partial(f_3 +_n g_3)/\partial z$$

Following by **(D1)** we can write

$$div(\mathbf{F} +_n \mathbf{G}) = div\mathbf{F} +_n div\mathbf{G} +_n \xi_{29}$$

where ξ_{29} is a random variable depends on n and m.

So, probability of equality

$$div(\mathbf{F} +_n \mathbf{G}) = div\mathbf{F} +_n div\mathbf{G}$$

is less than 1.

So, we proved

Theorem 5.17 *We have vector fields*

$$\boldsymbol{F} = (f_1(x,y,z), f_2(x,y,z), f_3(x,y,z))$$

and

$$\boldsymbol{G} = (g_1(x,y,z), g_2(x,y,z), g_3(x,y,z))$$

with $f_1, f_2, f_3, g_1, g_2, g_3, x, y, z$ belong to W_n. We assume that $f_1, f_2, f_3, g_1, g_2, g_3$ are differentiable functions, and all partial derivatives belong to W_n. Then

$$div(\boldsymbol{F} +_n \boldsymbol{G}) = div\boldsymbol{F} +_n div\boldsymbol{G} +_n \xi_{29}$$

Corollary 5.18 *The probability of equality*

$$div(\boldsymbol{F} +_n \boldsymbol{G}) = div\boldsymbol{F} +_n div\boldsymbol{G}$$

is less than 1.

(D10)

$$\boldsymbol{F} = (f_1(x, y, z), f_2(x, y, z), f_3(x, y, z))$$

with f_1, f_2, f_3, x, y, z belong to W_n. We assume that f_1, f_2, f_3 are differentiable functions, and all partial derivatives belong to W_n.

We consider now

$$div\mathbf{rot}\mathbf{F} =$$

$$= \partial(\partial f_3/\partial y -_n \partial f_2/\partial z)/\partial x +_n \partial(\partial f_3/\partial x -_n \partial f_1/\partial z)/\partial y$$
$$+_n \partial(\partial f_2/\partial x -_n \partial f_1/\partial y)/\partial z$$

Following by **(D1)** and **(D4)** we get

$$div\mathbf{rot}\mathbf{F} = \xi_{30}$$

where ξ_{30} is a random variable depends on n and m.

So, probability of equality

$$div\mathbf{rot}\mathbf{F} = 0$$

is less than 1.

So, we proved

Theorem 5.19 *We have vector field*

$$\boldsymbol{F} = (f_1(x, y, z), f_2(x, y, z), f_3(x, y, z))$$

with f_1, f_2, f_3, x, y, z belong to W_n. We assume that f_1, f_2, f_3 are differentiable functions, and all partial derivatives belong to W_n. Then

$$div\boldsymbol{rot}\boldsymbol{F} = \xi_{30}$$

Corollary 5.20 *The probability of equality*

$$div\boldsymbol{rot}\boldsymbol{F} = 0$$

is less than 1.

(D11)

We consider now "the chain rule" in Mathematics with Observers. Let's

$$f = f(u), u = g(v)$$

with f, g, u, v belong to W_n. We assume that f, g are differentiable functions, and all derivatives belong to W_n.

We consider now

$$F(v) = f(g(v))$$

We have

$$(f(g(v +_n \Delta v)) -_n f(g(v))) \times_n (\frac{1}{\Delta v}) = F'(v) = (f(g(v)))'$$

And

$$(f(g(v +_n \Delta v)) -_n f(g(v))) \times_n (\frac{1}{\Delta v}) =$$

$$= (f(g(v) +_n g(v +_n \Delta v) -_n g(v)) -_n f(g(v))) \times_n (\frac{1}{\Delta v}) =$$

$$= (f'(g(v)) \times_n (g(v +_n \Delta v) -_n g(v))) \times_n (\frac{1}{\Delta v}) =$$

$$= f'(g(v)) \times_n ((g(v +_n \Delta v) -_n g(v)) \times_n (\frac{1}{\Delta v})) +_n \xi_{49}$$

And

$$(f(g(v)))' = f'(g(v)) \times_n g'(v) +_n \xi_{49}$$

where ξ_{49} is a random variable depends on n and m.

So, probability of equality

$$(f(g(v)))' = f'(g(v)) \times_n g'(v)$$

is less than 1.

So, we proved

Theorem 5.21 *Let's*

$$f = f(u), u = g(v)$$

with f, g, u, v belong to W_n. We assume that f, g are differentiable functions, and all derivatives belong to W_n. Then

$$(f(g(v)))' = f'(g(v)) \times_n g'(v) +_n \xi_{49}$$

Corollary 5.22 *The probability of equality*

$$(f(g(v)))' = f'(g(v)) \times_n g'(v)$$

is less than 1.

(D12)
We have the following

Theorem 5.23 *Let's consider function*

$$y = f(u, v)$$

and let's

$$u = u(x)$$

and

$$v = v(x)$$

with f, x, u, v belong to W_n. We assume that f, u, v are differentiable functions, and all derivatives belong to W_n. Then

$$y'(x) = \partial y/\partial u \times_n du/dx +_n \partial y/\partial v \times_n dv/dx +_n \xi_{52}$$

where ξ_{52} is the random variable depends on n and m.

Corollary 5.24 *The probability of equality*

$$y'(x) = \partial y/\partial u \times_n du/dx +_n \partial y/\partial v \times_n dv/dx$$

is less than 1.

(D13)
Let's consider function

$$z = z(x, y)$$

with x, y, z belong to W_n. We assume that z is differentiable function, and all partial derivatives exist and belong to W_n.
We have

$$\Delta z = z(x +_n \Delta x, y +_n \Delta y) -_n z(x, y) = (z(x +_n \Delta x, y +_n \Delta y) -_n$$

$$-_n z(x +_n \Delta x, y)) +_n (z(x +_n \Delta x, y) -_n z(x, y)) =$$

$$= (z(x +_n \Delta x, y +_n \Delta y) -_n z(x +_n \Delta x, y)) \times_n \frac{\Delta y}{\Delta y} +_n$$

$$+_n (z(x +_n \Delta x, y) -_n z(x, y)) \times_n \frac{\Delta x}{\Delta x} =$$

$$= ((z(x +_n \Delta x, y +_n \Delta y) -_n z(x +_n \Delta x, y)) \times_n \frac{1}{\Delta y}) \times_n \Delta y +_n \xi_{53} +_n$$

$$+_n ((z(x +_n \Delta x, y) -_n z(x, y)) \times_n \frac{1}{\Delta x}) \times_n \Delta x +_n \xi_{54} =$$

$$= \partial z / \partial y (x, y) \times_n \Delta y +_n \partial z / \partial x (x, y) \times_n \Delta x +_n$$

$$+_n (\partial / \partial x (\partial z / \partial y)(x, y) \times_n \Delta x) \times_n \Delta y +_n \xi_{53} +_n \xi_{54} +_n \xi_{55} =$$

$$= \partial z / \partial x \times_n \Delta x +_n \partial z / \partial y \times_n \Delta y +_n \xi_{56}$$

where

$$\xi_{56} = (\partial / \partial x (\partial z / \partial y)(x, y) \times_n \Delta x) \times_n \Delta y +_n \xi_{53} +_n \xi_{54} +_n \xi_{55}$$

and ξ_{53}, ξ_{54} and ξ_{55} are random variables depend on n and m.
Also note that the numbers $\frac{1}{\Delta x}$, $\frac{1}{\Delta y}$ exist with probability less than 1.
So, the probability of equality

$$\Delta z = \partial z / \partial x \times_n \Delta x +_n \partial z / \partial y \times_n \Delta y$$

is less than 1.
So, we proved

Theorem 5.25 *Let's consider function*

$$z = z(x, y)$$

with x, y, z belong to W_n. We assume that z is differentiable function, and all partial derivatives exist and belong to W_n. Then

$$\Delta z = \partial z / \partial x \times_n \Delta x +_n \partial z / \partial y \times_n \Delta y +_n \xi_{56}$$

Corollary 5.26 *The probability of equality*

$$\Delta z = \partial z / \partial x \times_n \Delta x +_n \partial z / \partial y \times_n \Delta y$$

is less than 1.

(D14)

Let's consider functions

$$z = z(x, y), w = w(u, v)$$

with x, y, z, u, v, w belong to W_n. We assume that z, w are differentiable functions, and all partial derivatives exist and belong to W_n.

By **(D2)** we can write

$$z(x +_n \Delta x, y +_n \Delta y) \times_n w(u +_n \Delta u, v +_n \Delta v) -_n z(x, y) \times_n w(u, v) =$$

$$= (z(x, y) +_n \Delta z +_n \xi_{57}) \times_n (w(u, v) +_n \Delta w +_n \xi_{58}) -_n z(x, y) \times_n w(u, v) =$$

$$= z(x, y) \times_n w(u, v) +_n z(x, y) \times_n \Delta w +_n z(x, y) \times_n \xi_{58} +_n \Delta z \times_n w(u, v) +_n$$

$$+_n \Delta z \times_n \Delta w +_n \Delta z \times_n \xi_{58} +_n$$

$$+_n \xi_{57} \times_n w(u, v) +_n \xi_{57} \times_n \Delta w +_n \xi_{57} \times_n \xi_{58} +_n \xi_{59} -_n z(x, y) \times_n w(u, v) =$$

$$= \Delta z \times_n w(u, v) +_n z(x, y) \times_n \Delta w +_n \xi_{60}$$

where $\xi_{57}, \xi_{58}, \xi_{59}$ and ξ_{60} are random variables depend on n and m. That means

$$\Delta(z \times_n w) = \Delta z \times_n w +_n z \times_n \Delta w +_n \xi_{60}$$

So, the probability of equality

$$\Delta(z \times_n w) = \Delta z \times_n w +_n z \times_n \Delta w$$

is less than 1.

So, we proved

Theorem 5.27 *Let's consider functions*

$$z = z(x, y), w = w(u, v)$$

with x, y, z, u, v, w belong to W_n. We assume that z, w are differentiable functions, and all partial derivatives exist and belong to W_n. Then

$$\Delta(z \times_n w) = \Delta z \times_n w +_n z \times_n \Delta w +_n \xi_{60}$$

Corollary 5.28 *The probability of equality*

$$\Delta(z \times_n w) = \Delta z \times_n w +_n z \times_n \Delta w$$

is less than 1.

(D15)
Let f be the differentiable function of several variables $x_1, \ldots, x_k$, and $f, x_1, \ldots, x_k \in W_n$, and all partial derivatives belong to W_n.
By **(D4)** we get

$$\partial/\partial x_j(\partial f/\partial x_i) = \partial/\partial x_i(\partial f/\partial x_j) +_n \xi_{i,j}$$

where

$$i, j \in (1, \ldots, k), i \neq j$$

and $\xi_{i,j}$ are the random variables depend on n and m.
Total we have $\frac{k \times_n (k-1)}{2}$ such relations.
So, the probability of equality

$$\partial/\partial x_j(\partial f/\partial x_i) = \partial/\partial x_i(\partial f/\partial x_j)$$

is less than 1.
So, we proved

Theorem 5.29 *Let f be the differentiable function of several variables $x_1, \ldots, x_k$, and $f, x_1, \ldots, x_k \in W_n$, and all partial derivatives belong to W_n. Then we get*

$$\partial/\partial x_j(\partial f/\partial x_i) = \partial/\partial x_i(\partial f/\partial x_j) +_n \xi_{i,j}$$

where

$$i, j \in (1, \ldots, k), i \neq j$$

and $\xi_{i,j}$ are the random variables depend on n and m. Total we have $\frac{k \times_n (k-1)}{2}$ such relations.

Corollary 5.30 *The probability of equality*

$$\partial/\partial x_j(\partial f/\partial x_i) = \partial/\partial x_i(\partial f/\partial x_j)$$

is less than 1.

(D16)
Let's consider function

$$z = z(x, y)$$

with x, y, z belong to W_n. We assume that z is differentiable function, and all partial derivatives exist and belong to W_n.
By **(D13)** we have

$$\Delta z = \partial z / \partial x \times_n \Delta x +_n \partial z / \partial y \times_n \Delta y +_n \xi_{56}$$

Let's now we have

$$\Delta z = M \times_n \Delta x +_n N \times_n \Delta y$$

where

$$M = M(x, y), N = N(x, y)$$

We get

$$\partial z / \partial x \times_n \Delta x +_n \partial z / \partial y \times_n \Delta y +_n \xi_{56} = M \times_n \Delta x +_n N \times_n \Delta y$$

i.e.

$$(\partial z / \partial x \times_n \Delta x -_n M \times_n \Delta x) +_n (\partial z / \partial y \times_n \Delta y -_n N \times_n \Delta y) = \xi_{56}$$

$$(\partial z / \partial x -_n M) \times_n \Delta x +_n \xi_{61} +_n (\partial z / \partial y -_n N) \times_n \Delta y +_n \xi_{62} = \xi_{56}$$

where ξ_{61}, ξ_{62} are random variables depend on n and m.
And we get

$$M = \partial z / \partial x +_n \xi_{63}$$

$$N = \partial z / \partial y +_n \xi_{64}$$

where ξ_{63}, ξ_{64} are random variables depend on n and m.
So, the probabilities of equalities

$$M = \partial z / \partial x$$

$$N = \partial z / \partial y$$

are less than 1.
So, we proved

Theorem 5.31 *Let's consider function*

$$z = z(x, y)$$

with x, y, z belong to W_n. We assume that z is differentiable function, and all partial derivatives exist and belong to W_n. Let's now we have

$$\Delta z = M \times_n \Delta x +_n N \times_n \Delta y$$

where
$$M = M(x, y), N = N(x, y)$$

Then we get
$$M = \partial z/\partial x +_n \xi_{63}$$
$$N = \partial z/\partial y +_n \xi_{64}$$

Corollary 5.32 *The probabilities of equalities*
$$M = \partial z/\partial x$$
$$N = \partial z/\partial y$$
are less than 1.

(D17)
Let's $\mathbf{a} = (a_1, a_2, a_3), \mathbf{b} = (b_1, b_2, b_3) \in E_3 W_n$, $\alpha \in W_n$ depend on $t \in W_n$, and (') means derivative by t.

We have the following statements:

((D17a))

$$(\mathbf{a} +_n \mathbf{b})' = \mathbf{a}' +_n \mathbf{b}' +_n \xi_{65}$$

where ξ_{65} is random variable depends on n and m.
So, the probability of equality

$$(\mathbf{a} +_n \mathbf{b})' = \mathbf{a}' +_n \mathbf{b}'$$

is less than 1.
So, we proved

Theorem 5.33 *Let's* $a = (a_1, a_2, a_3), b = (b_1, b_2, b_3) \in E_3 W_n$ *depend on* $t \in W_n$, *and (') means derivative by* t. *Then*

$$(a +_n b)' = a' +_n b' +_n \xi_{65}$$

Corollary 5.34 *The probability of equality*
$$(a +_n b)' = a' +_n b'$$
is less than 1.

((D17b))

$$(\alpha \times_n \mathbf{a})' = \alpha' \times_n \mathbf{a} +_n \alpha \times_n \mathbf{a}' +_n \xi_{66}$$

where ξ_{66} is random variable depends on n and m.
So, the probabilities of equality

$$(\alpha \times_n \mathbf{a})' = \alpha' \times_n \mathbf{a} +_n \alpha \times_n \mathbf{a}'$$

is less than 1.
So, we proved

Theorem 5.35 *Let's* $a = (a_1, a_2, a_3), b = (b_1, b_2, b_3) \in E_3 W_n$, $\alpha \in W_n$ *depend on* $t \in W_n$, *and* (') *means derivative by* t. *Then*

$$(\alpha \times_n a)' = \alpha' \times_n a +_n \alpha \times_n a' +_n \xi_{66}$$

Corollary 5.36 *The probability of equality*

$$(\alpha \times_n a)' = \alpha' \times_n a +_n \alpha \times_n a'$$

is less than 1.

(D17c)

$$(\mathbf{a}, \mathbf{b})' = (\mathbf{a}', \mathbf{b}) +_n (\mathbf{a}, \mathbf{b}') +_n \xi_{67}$$

where ξ_{67} is random variable depends on n and m.
So, the probabilities of equality

$$(\mathbf{a}, \mathbf{b})' = (\mathbf{a}', \mathbf{b}) +_n (\mathbf{a}, \mathbf{b}')$$

is less than 1.
So, we proved

Theorem 5.37 *Let's* $a = (a_1, a_2, a_3), b = (b_1, b_2, b_3) \in E_3 W_n$ *depend on* $t \in W_n$, *and* (') *means derivative by* t. *Then*

$$(a, b)' = (a', b) +_n (a, b') +_n \xi_{67}$$

Corollary 5.38 *The probability of equality*

$$(a, b)' = (a', b) +_n (a, b')$$

is less than 1.

(D17d)

$$(\mathbf{a} \times \mathbf{b})' = \mathbf{a}' \times \mathbf{b} +_n \mathbf{a} \times \mathbf{b}' +_n \xi_{68}$$

where ξ_{68} is random variable depends on n and m.
So, the probability of equality

$$(\mathbf{a} \times \mathbf{b})' = \mathbf{a}' \times \mathbf{b} +_n \mathbf{a} \times \mathbf{b}'$$

is less than 1.
So, we proved

Theorem 5.39 *Let's* $a = (a_1, a_2, a_3), b = (b_1, b_2, b_3) \in E_3 W_n$ *depend on* $t \in W_n$, *and* (') *means derivative by* t. *Then*

$$(a \times b)' = a' \times b +_n a \times b' +_n \xi_{68}$$

Corollary 5.40 *The probability of equality*

$$(a \times b)' = a' \times b +_n a \times b'$$

is less than 1.

(D18)
Let's

$$\alpha = \alpha(x, y, z)$$

is a scalar function,

$$\mathbf{F} = (f_1(x, y, z), f_2(x, y, z), f_3(x, y, z))$$

is a vector field
   with $\alpha, f_1, f_2, f_3, x, y, z$ belong to W_n. We assume that α, f_1, f_2, f_3 are differentiable functions, and all partial derivatives belong to W_n.
   We consider now

$$div(\alpha \times_n \mathbf{F}) =$$

$$= (\nabla, \alpha \times_n \mathbf{F}) = \partial(\alpha \times_n f_1)/\partial x +_n \partial(\alpha \times_n f_2)/\partial y +_n \partial(\alpha \times_n f_3)/\partial z =$$

$$= \partial\alpha/\partial x \times_n f_1 +_n \alpha \times_n \partial f_1/\partial x +_n \xi_{69} +_n \partial\alpha/\partial y \times_n f_2 +_n \alpha \times_n \partial f_2/\partial y +_n \xi_{70} +_n$$

$$+_n \partial\alpha/\partial z \times_n f_3 +_n \alpha \times_n \partial f_3/\partial z +_n \xi_{71} =$$

$$= \alpha \times_n div\mathbf{F} +_n \xi_{72} +_n (\mathbf{F}, \mathbf{grad}\alpha) +_n \xi_{69} +_n \xi_{70} +_n \xi_{71}$$

where $\xi_{69}, \xi_{70}, \xi_{71}, \xi_{72}$ are the random variables depend on n and m.
So, we get

$$div(\alpha \times_n \mathbf{F}) = \alpha \times_n div\mathbf{F} +_n (\mathbf{F}, \mathbf{grad}\alpha) +_n \xi_{73}$$

where ξ_{73} is random variable depends on n and m.
So, the probability of equality

$$div(\alpha \times_n \mathbf{F}) = \alpha \times_n div\mathbf{F} +_n (\mathbf{F}, \mathbf{grad}\alpha)$$

is less than 1.
So, we proved

Theorem 5.41 *Let's*

$$\alpha = \alpha(x, y, z)$$

is a scalar function,

$$\boldsymbol{F} = (f_1(x, y, z), f_2(x, y, z), f_3(x, y, z))$$

is a vector field with $\alpha, f_1, f_2, f_3, x, y, z$ *belong to* W_n. *We assume that* α, f_1, f_2, f_3 *are differentiable functions, and all partial derivatives belong to* W_n. *Then*

$$div(\alpha \times_n \boldsymbol{F}) = \alpha \times_n div\boldsymbol{F} +_n (\boldsymbol{F}, \boldsymbol{grad}\alpha) +_n \xi_{73}$$

Corollary 5.42 *The probability of equality*

$$div(\alpha \times_n \boldsymbol{F}) = \alpha \times_n div\boldsymbol{F} +_n (\boldsymbol{F}, \boldsymbol{grad}\alpha)$$

is less than 1.

(D19)
Let's

$$\boldsymbol{F} = (f_1(x, y, z), f_2(x, y, z), f_3(x, y, z))$$

is a vector field with f_1, f_2, f_3, x, y, z belong to W_n. We assume that f_1, f_2, f_3 are differentiable functions, and all partial derivatives belong to W_n. We consider now

$$\nabla = (\partial/\partial x, \partial/\partial y, \partial/\partial z)$$

$$(\nabla, \mathbf{F}) = \partial f_1/\partial x +_n \partial f_2/\partial y +_n \partial f_3/\partial z$$

$$\nabla(\nabla, \mathbf{F}) = (\partial/\partial x(\partial f_1/\partial x +_n \partial f_2/\partial y +_n \partial f_3/\partial z),$$

$$\partial/\partial y(\partial f_1/\partial x +_n \partial f_2/\partial y +_n \partial f_3/\partial z),$$

$$\partial/\partial z(\partial f_1/\partial x +_n \partial f_2/\partial y +_n \partial f_3/\partial z)) =$$

$$= (\partial^2 f_1/\partial x^2 +_n \partial^2 f_2/\partial y\partial x +_n \partial^2 f_3/\partial z\partial x +_n \xi_{74},$$

$$\partial^2 f_1/\partial x\partial y +_n \partial^2 f_2/\partial y^2 +_n \partial^2 f_3/\partial z\partial y +_n \xi_{75},$$

$$\partial^2 f_1/\partial x\partial z +_n \partial^2 f_2/\partial y\partial z +_n \partial^2 f_3/\partial z^2 +_n \xi_{76}) =$$

$$= (\partial^2 f_1/\partial x^2 +_n \partial^2 f_2/\partial x\partial y +_n \partial^2 f_3/\partial x\partial z +_n \xi_{74} +_n \xi_{77} +_n \xi_{78},$$

$$\partial^2 f_1/\partial x \partial y +_n \partial^2 f_2/\partial y^2 +_n \partial^2 f_3/\partial y \partial z +_n \xi_{75} +_n \xi_{79},$$

$$\partial^2 f_1/\partial x \partial z +_n \partial^2 f_2/\partial y \partial z +_n \partial^2 f_3/\partial z^2 +_n \xi_{76})$$

where

$$\xi_{74}, \xi_{75}, \xi_{76}, \xi_{77}, \xi_{78}, \xi_{79}$$

are the random variables depend on n and m.
Let's consider now

$$\nabla \times \mathbf{F} = (\partial f_3/\partial y -_n \partial f_2/\partial z, \partial f_1/\partial z -_n \partial f_3/\partial x, \partial f_2/\partial x -_n \partial f_1/\partial y)$$

$$\nabla \times (\nabla \times \mathbf{F}) = (\partial/\partial y (\partial f_2/\partial x -_n \partial f_1/\partial y) -_n \partial/\partial z (\partial f_1/\partial z -_n \partial f_3/\partial x),$$

$$\partial/\partial z (\partial f_3/\partial y -_n \partial f_2/\partial z) -_n \partial/\partial x (\partial f_2/\partial x -_n \partial f_1/\partial y),$$

$$\partial/\partial x (\partial f_1/\partial z -_n \partial f_3/\partial x) -_n \partial/\partial y (\partial f_3/\partial y -_n \partial f_2/\partial z)) =$$

$$= (\partial^2 f_2/\partial x \partial y -_n \partial^2 f_1/\partial y^2 +_n \xi_{80} -_n \partial^2 f_1/\partial z^2 +_n \partial^2 f_3/\partial x \partial z +_n \xi_{81},$$

$$\partial^2 f_3/\partial y \partial z -_n \partial^2 f_2/\partial z^2 +_n \xi_{82} -_n \partial^2 f_2/\partial x^2 +_n \partial^2 f_1/\partial y \partial x +_{83},$$

$$\partial^2 f_1/\partial z \partial x -_n \partial^2 f_3/\partial x^2 +_n \xi_{84} -_n \partial^2 f_3/\partial y^2 +_n \partial^2 f_2/\partial z \partial y +_n \xi_{85}) =$$

$$= (\partial^2 f_2/\partial x \partial y -_n \partial^2 f_1/\partial y^2 +_n \xi_{80} -_n \partial^2 f_1/\partial z^2 +_n \partial^2 f_3/\partial x \partial z +_n \xi_{81},$$

$$\partial^2 f_3/\partial y \partial z -_n \partial^2 f_2/\partial z^2 +_n \xi_{82} -_n \partial^2 f_2/\partial x^2 +_n \partial^2 f_1/\partial x \partial y +_n \xi_{86} +_n \xi_{83},$$

$$\partial^2 f_1/\partial x \partial z +_n \xi_{87} -_n \partial^2 f_3/\partial x^2 +_n \xi_{84} -_n \partial^2 f_3/\partial y^2 +_n \partial^2 f_2/\partial y \partial z +_n \xi_{88} +_n \xi_{85})$$

where

$$\xi_{80}, \xi_{81}, \xi_{82}, \xi_{83}, \xi_{84}, \xi_{85}, \xi_{86}, \xi_{87}, \xi_{88}$$

are the random variables depend on n and m.

Let's consider now

$$\nabla(\nabla, \mathbf{F}) -_n \nabla \times (\nabla \times \mathbf{F}) =$$

$$= (\partial^2 f_1/\partial x^2 +_n \partial^2 f_2/\partial x \partial y +_n \partial^2 f_3/\partial x \partial z +_n \xi_{74} +_n \xi_{77} +_n \xi_{78} -_n$$

$$-_n (\partial^2 f_2/\partial x \partial y -_n \partial^2 f_1/\partial y^2 +_n \xi_{80} -_n \partial^2 f_1/\partial z^2 +_n \partial^2 f_3/\partial x \partial z +_n \xi_{81}),$$

$$\partial^2 f_1/\partial x \partial y +_n \partial^2 f_2/\partial y^2 +_n \partial^2 f_3/\partial y \partial z +_n \xi_{75} +_n \xi_{79} -_n$$

$$-_n (\partial^2 f_3/\partial y \partial z -_n \partial^2 f_2/\partial z^2 +_n \xi_{82} -_n \partial^2 f_2/\partial x^2 +_n \partial^2 f_1/\partial x \partial y +_n \xi_{86} +_n \xi_{83}),$$

$$\partial^2 f_1/\partial x \partial z +_n \partial^2 f_2/\partial y \partial z +_n \partial^2 f_3/\partial z^2 +_n \xi_{76} -_n$$

$$-_n (\partial^2 f_1/\partial x \partial z +_n \xi_{87} -_n \partial^2 f_3/\partial x^2 +_n \xi_{84} -_n \partial^2 f_3/\partial y^2$$
$$+_n \partial^2 f_2/\partial y \partial z +_n \xi_{88} +_n \xi_{85})) =$$

$$= (\partial^2 f_1/\partial x^2 +_n \partial^2 f_1/\partial y^2 +_n \partial^2 f_1/\partial z^2 +_n \xi_{74} +_n \xi_{77} +_n \xi_{78} -_n \xi_{80} -_n \xi_{81},$$

$$\partial^2 f_2/\partial y^2 +_n \partial^2 f_2/\partial z^2 +_n \partial^2 f_2/\partial x^2 +_n \xi_{75} +_n \xi_{79} -_n \xi_{82} -_n \xi_{86} -_n \xi_{83},$$

$$\partial^2 f_3/\partial z^2 +_n \partial^2 f_3/\partial x^2 +_n \partial^2 f_3/\partial y^2 +_n \xi_{76} -_n \xi_{87} -_n \xi_{84} -_n \xi_{88} -_n \xi_{85})$$

Let's consider now Laplacian $\Delta \mathbf{F}$:

$$(\nabla, \nabla)\mathbf{F} = \Delta \mathbf{F} = (\partial^2 f_1/\partial x^2 +_n \partial^2 f_1/\partial y^2 +_n \partial^2 f_1/\partial z^2,$$

$$\partial^2 f_2/\partial x^2 +_n \partial^2 f_2/\partial y^2 +_n \partial^2 f_2/\partial z^2,$$

$$\partial^2 f_3/\partial x^2 +_n \partial^2 f_3/\partial y^2 +_n \partial^2 f_3/\partial z^2)$$

So, we get

$$\nabla(\nabla, \mathbf{F}) -_n \nabla \times (\nabla \times \mathbf{F}) = \Delta \mathbf{F} +_n \xi_{89}$$

where ξ_{89} is a random vector depends on n and m:

$$\xi_{89} = (\xi_{74} +_n \xi_{77} +_n \xi_{78} -_n \xi_{80} -_n \xi_{81}, \xi_{75} +_n \xi_{79} -_n \xi_{82} -_n \xi_{86} -_n \xi_{83}, \xi_{76} -_n \xi_{87} -_n$$

$$-_n \xi_{84} -_n \xi_{88} -_n \xi_{85})$$

So, the probability of equality

$$\nabla(\nabla, \mathbf{F}) -_n \nabla \times (\nabla \times \mathbf{F}) = \Delta \mathbf{F}$$

is less than 1.
So, we proved

Theorem 5.43 *Let's*

$$\boldsymbol{F} = (f_1(x, y, z), f_2(x, y, z), f_3(x, y, z))$$

is a vector field with f_1, f_2, f_3, x, y, z belong to W_n. We assume that f_1, f_2, f_3 are differentiable functions, and all partial derivatives belong to W_n. Then

$$\nabla(\nabla, \boldsymbol{F}) -_n \nabla \times (\nabla \times \boldsymbol{F}) = \Delta \boldsymbol{F} +_n \xi_{89}$$

Corollary 5.44 *The probability of equality*

$$\nabla(\nabla, \boldsymbol{F}) -_n \nabla \times (\nabla \times \boldsymbol{F}) = \Delta \boldsymbol{F}$$

is less than 1.

(D20)
By **D10** we have

$$div(\nabla \times (\nabla \times \mathbf{F})) = \xi_{90}$$

and by **D8** we have

$$\mathbf{rot}(\nabla(\nabla, \mathbf{F})) = \xi_{91}$$

where ξ_{90} is a random variable depends on n and m, and ξ_{91} is a random vector
depends on n and m.
By **D9** and **D19** we have

$$div(\Delta \mathbf{F}) = div(\nabla(\nabla, \mathbf{F})) -_n div(\nabla \times (\nabla \times \mathbf{F})) +_n \xi_{92} = div(\nabla(\nabla, \mathbf{F})) -_n \xi_{90} +_n \xi_{92}$$

and by **D5** and **D19** we have

$$\mathbf{rot}(\Delta \mathbf{F}) = \mathbf{rot}(\nabla(\nabla, \mathbf{F})) -_n \mathbf{rot}(\nabla \times (\nabla \times \mathbf{F})) +_n \xi_{93}$$
$$= \xi_{91} -_n \mathbf{rot}(\nabla \times (\nabla \times \mathbf{F})) +_n \xi_{93}$$

where ξ_{92} is a random variable depends on n and m, and ξ_{93} is a random vector
depends on n and m.
So, we proved

Theorem 5.45 *Let's*

$$\mathbf{F} = (f_1(x, y, z), f_2(x, y, z), f_3(x, y, z))$$

is a vector field with f_1, f_2, f_3, x, y, z belong to W_n. We assume that f_1, f_2, f_3 are differentiable functions, and all partial derivatives belong to W_n. Then

$$div(\Delta \mathbf{F}) = div(\nabla(\nabla, \mathbf{F})) -_n div(\nabla \times (\nabla \times \mathbf{F})) +_n \xi_{92} = div(\nabla(\nabla, \mathbf{F})) -_n \xi_{90} +_n \xi_{92}$$

$$\mathbf{rot}(\Delta \mathbf{F}) = \mathbf{rot}(\nabla(\nabla, \mathbf{F})) -_n \mathbf{rot}(\nabla \times (\nabla \times \mathbf{F})) +_n \xi_{93}$$
$$= \xi_{91} -_n \mathbf{rot}(\nabla \times (\nabla \times \mathbf{F})) +_n \xi_{93}$$

Corollary 5.46 *The probabilities of equalities*

$$div(\Delta \mathbf{F}) = div(\nabla(\nabla, \mathbf{F})) -_n div(\nabla \times (\nabla \times \mathbf{F})) = div(\nabla(\nabla, \mathbf{F}))$$

$$\mathbf{rot}(\Delta \mathbf{F}) = \mathbf{rot}(\nabla(\nabla, \mathbf{F})) -_n \mathbf{rot}(\nabla \times (\nabla \times \mathbf{F})) = -_n \mathbf{rot}(\nabla \times (\nabla \times \mathbf{F}))$$

are less than 1.

(D21)
For vector fields $\mathbf{a}, \mathbf{b} \in E_3 W_n$ let's consider and compare two functions:

$$div(\mathbf{a} \times \mathbf{b})$$

and

$$(\mathbf{b}, \mathbf{rota}) -_n (\mathbf{a}, \mathbf{rotb})$$

Let's

$$\mathbf{a} = (a_1(x, y, z), a_2(x, y, z), a_3(x, y, z))$$

$$\mathbf{b} = (b_1(x, y, z), b_2(x, y, z), b_3(x, y, z))$$

Then

$$\mathbf{a} \times \mathbf{b} = (a_2 \times_n b_3 -_n a_3 \times_n b_2, a_3 \times_n b_1 -_n a_1 \times_n b_3, a_1 \times_n b_2 -_n a_2 \times_n b_1)$$

and

$$div(\mathbf{a} \times \mathbf{b}) = \partial(a_2 \times_n b_3 -_n a_3 \times_n b_2)/\partial x +_n \partial(a_3 \times_n b_1 -_n a_1 \times_n b_3)/\partial y +_n$$

$$+_n \partial(a_1 \times_n b_2 -_n a_2 \times_n b_1)/\partial z =$$

$$= \partial a_2/\partial x \times_n b_3 +_n a_2 \times_n \partial b_3/\partial x -_n \partial a_3/\partial x \times_n b_2 -_n a_3 \times_n \partial b_2/\partial x +_n \xi_{101} +_n$$

$$+_n \partial a_3/\partial y \times_n b_1 +_n a_3 \times_n \partial b_1/\partial y -_n \partial a_1/\partial y \times_n b_3 -_n a_1 \times_n \partial b_3/\partial y +_n \xi_{102} +_n$$

$$+_n \partial a_1/\partial z \times_n b_2 +_n a_1 \times_n \partial b_2/\partial z -_n \partial a_2/\partial z \times_n b_1 -_n a_2 \times_n \partial b_1/\partial z +_n \xi_{103}$$

where $\xi_{101}, \xi_{102}, \xi_{103}$ are the random variables depend on n and m.
Now we get

$$(\mathbf{b}, \mathbf{rota}) = (\mathbf{b}, (\nabla \times \mathbf{a})) =$$

$$= (\mathbf{b}, (\partial a_3/\partial y -_n \partial a_2/\partial z) \times_n \mathbf{i} -_n (\partial a_3/\partial x -_n \partial a_1/\partial z) \times_n \mathbf{j} +_n (\partial a_2/\partial x$$
$$-_n \partial a_1/\partial y) \times_n \mathbf{k}) =$$

$$= b_1 \times_n (\partial a_3/\partial y -_n \partial a_2/\partial z) -_n b_2 \times_n (\partial a_3/\partial x -_n \partial a_1/\partial z)$$
$$+_n b_3 \times_n (\partial a_2/\partial x -_n \partial a_1/\partial y) =$$

$$= b_1 \times_n \partial a_3/\partial y - b_1 \times_n \partial a_2/\partial z +_n \xi_{104} -_n$$

$$-_n b_2 \times_n \partial a_3/\partial x +_n b_2 \times_n \partial a_1/\partial z +_n \xi_{105} +_n$$

$$+_n b_3 \times_n \partial a_2/\partial x - b_3 \times_n \partial a_1/\partial y +_n \xi_{106}$$

where $\xi_{104}, \xi_{105}, \xi_{106}$ are the random variables depend on n and m.
And finally let's consider

$$(\mathbf{a}, \mathbf{rotb}) = (\mathbf{a}, (\nabla \times \mathbf{b})) =$$

$$= (\mathbf{a}, (\partial b_3/\partial y -_n \partial b_2/\partial z) \times_n \mathbf{i} -_n (\partial b_3/\partial x -_n \partial b_1/\partial z) \times_n \mathbf{j}$$
$$+_n (\partial b_2/\partial x -_n \partial b_1/\partial y) \times_n \mathbf{k}) =$$

$$= a_1 \times_n (\partial b_3/\partial y -_n \partial b_2/\partial z) -_n a_2 \times_n (\partial b_3/\partial x -_n \partial b_1/\partial z)$$
$$+_n a_3 \times_n (\partial b_2/\partial x -_n \partial b_1/\partial y) =$$

$$= a_1 \times_n \partial b_3/\partial y - a_1 \times_n \partial b_2/\partial z +_n \xi_{107}-_n$$

$$-_n a_2 \times_n \partial b_3/\partial x +_n a_2 \times_n \partial b_1/\partial z +_n \xi_{108}+_n$$

$$+_n a_3 \times_n \partial b_2/\partial x - a_3 \times_n \partial b_1/\partial y +_n \xi_{109}$$

where $\xi_{107}, \xi_{108}, \xi_{109}$ are the random variables depend on n and m. So, we have

$$div(\mathbf{a} \times \mathbf{b}) = (\mathbf{b}, \mathbf{rota}) -_n (\mathbf{a}, \mathbf{rotb}) +_n \xi_{110}$$

where ξ_{110} is the random variable depends on n and m. And probability of equality

$$div(\mathbf{a} \times \mathbf{b}) = (\mathbf{b}, \mathbf{rota}) -_n (\mathbf{a}, \mathbf{rotb})$$

is less than 1.
So, we proved

Theorem 5.47 *Let's vector fields $a, b \in E_3W_n$. We assume that their coordinates are differentiable functions, and all partial derivatives belong to W_n. Then*
$$div(a \times b) = (b, rota) -_n (a, rotb) +_n \xi_{110}$$

Corollary 5.48 *The probability of equality*
$$div(a \times b) = (b, rota) -_n (a, rotb)$$

is less than 1.

Let's consider now the Integral properties in Mathematics with Observers.
I1.
Let's
$$f(x) = c, c = const$$

and
$$a, b, x, \Delta x, k, c, (b -_n a) \times_n c \in W_n$$

We get

$$\int_a^b c \times_n \Delta x = \sum_{i=1}^k {}^n c \times_n \Delta x = c \times_n \sum_{i=1}^k {}^n \Delta x +_n \iota_1 = (b -_n a) \times_n c +_n \iota_1$$

where ι_1 is a random variable depends on n and m.
So, the probability of equality

$$\int_a^b c \times_n \Delta x = (b -_n a) \times_n c$$

is less than 1.
So, we proved

Theorem 5.49 *Let's*

$$f(x) = c, c = const$$

and

$$a, b, x, \Delta x, k, c, (b -_n a) \times_n c \in W_n$$

Then

$$\int_a^b c \times_n \Delta x = \sum_{i=1}^k {}^n c \times_n \Delta x = c \times_n \sum_{i=1}^k {}^n \Delta x$$
$$+_n \iota_1 = (b -_n a) \times_n c +_n \iota_1$$

Corollary 5.50 *The probability of equality*

$$\int_a^b c \times_n \Delta x = (b -_n a) \times_n c$$

is less than 1.

I2.
Let's

$$f(x) = f_1(x) +_n f_2(x)$$

and

$$a, b, x, \Delta x, k, f_1(x), f_2(x), f(x) \in W_n$$

We get

$$\int_a^b f(x) \times_n \Delta x = \int_a^b (f_1(x) +_n f_2(x)) \times_n \Delta x =$$

$$= \sum_{i=1}^k {}^n (f_1(x_i) +_n f_2(x_i)) \times_n \Delta x =$$

$$= \sum_{i=1}^k {}^n f_1(x_i) \times_n \Delta x +_n \sum_{i=1}^k {}^n f_2(x_i) \times_n \Delta x +_n \iota_2 =$$

$$= \int_a^b f_1(x) \times_n \Delta x +_n \int_a^b f_2(x) \times_n \Delta x +_n \iota_2$$

where ι_2 is a random variable depends on n and m.
So, the probability of equality

$$\int_a^b f(x) \times_n \Delta x = \int_a^b f_1(x) \times_n \Delta x +_n \int_a^b f_2(x) \times_n \Delta x$$

is less than 1.
So, we proved

Theorem 5.51 *Let's*
$$f(x) = f_1(x) +_n f_2(x)$$

and
$$a, b, x, \Delta x, k, f_1(x), f_2(x), f(x) \in W_n$$

Then

$$\int_a^b f(x) \times_n \Delta x = \int_a^b (f_1(x) +_n f_2(x)) \times_n \Delta x = \int_a^b f_1(x) \times_n \Delta x$$

$$+_n \int_a^b f_2(x) \times_n \Delta x +_n \iota_2$$

Corollary 5.52 *The probability of equality*

$$\int_a^b f(x) \times_n \Delta x = \int_a^b f_1(x) \times_n \Delta x +_n \int_a^b f_2(x) \times_n \Delta x$$

is less than 1.

I3.
Let's

$$F(x) = c \times_n f(x), c = const$$

and

$$a, b, c, x, \Delta x, k, f(x), c \times_n f(x) \in W_n$$

We get

$$\int_a^b F(x) \times_n \Delta x = \int_a^b (c \times_n f(x)) \times_n \Delta x =$$

$$= \sum_{i=1}^k {}^n (c \times_n f(x_i)) \times_n \Delta x = \sum_{i=1}^k {}^n c \times_n (f(x_i) \times_n \Delta x) +_n \iota_3 =$$

$$= c \times_n \sum_{i=1}^{k} {}^n f(x_i) \times_n \Delta x +_n \iota_4 +_n \iota_3 =$$

$$= c \times_n \int_a^b f(x) \times_n \Delta x +_n \iota_4 +_n \iota_3$$

where ι_3, ι_4 are random variables depend on n and m.
So, the probabilities of equality

$$\int_a^b (c \times_n f(x)) \times_n \Delta x = c \times_n \int_a^b f(x) \times_n \Delta x$$

is less than 1.
So, we proved

Theorem 5.53 *Let's*

$$F(x) = c \times_n f(x), c = const$$

and

$$a, b, c, x, \Delta x, k, f(x), c \times_n f(x) \in W_n$$

Then

$$\int_a^b F(x) \times_n \Delta x = \int_a^b (c \times_n f(x)) \times_n \Delta x = c \times_n \int_a^b f(x) \times_n \Delta x +_n \iota_4 +_n \iota_3$$

Corollary 5.54 *The probability of equality*

$$\int_a^b (c \times_n f(x)) \times_n \Delta x = c \times_n \int_a^b f(x) \times_n \Delta x$$

is less than 1.

I4. Let's

$$a < b_1 < b$$

and

$$a, b_1, b, x, \Delta x, k, f(x), \int_a^b f(x) \times_n \Delta x, \int_a^{b_1} f(x) \times_n \Delta x, \int_{b_1}^b f(x) \times_n \Delta x,$$

$$\int_a^{b_1} f(x) \times_n \Delta x +_n \int_{b_1}^b f(x) \times_n \Delta x \in W_n$$

If $x_{p+n1} = b_1$ for some positive integer $p < k$, we get

$$\int_a^b f(x) \times_n \Delta x = \sum_{i=1}^k {}^n f(x_i) \times_n \Delta x =$$

$$= \sum_{i=1}^p {}^n f(x_i) \times_n \Delta x +_n \sum_{i=p+_n 1}^k {}^n f(x_i) \times_n \Delta x =$$

$$= \int_a^{b_1} f(x) \times_n \Delta x +_n \int_{b_1}^b f(x) \times_n \Delta x$$

Let's note the value of Δx is an integral parameter. The main value of parameter Δx is $\Delta x_{min} = 0.\underbrace{0...0}_{n-1} 2$.

So, for given n we have limiting on numbers a, b_1, b, p, k.

So, we proved

Theorem 5.55 *Let's*

$$a < b_1 < b$$

and

$$a, b_1, b, x, \Delta x, k, f(x), \int_a^b f(x) \times_n \Delta x, \int_a^{b_1} f(x) \times_n \Delta x, \int_{b_1}^b f(x) \times_n \Delta x,$$

$$\int_a^{b_1} f(x) \times_n \Delta x +_n \int_{b_1}^b f(x) \times_n \Delta x \in W_n$$

Let's $x_{p+_n 1} = b_1$ for some positive integer $p < k$. Then

$$\int_a^b f(x) \times_n \Delta x = \sum_{i=1}^k {}^n f(x_i) \times_n \Delta x = \int_a^{b_1} f(x) \times_n \Delta x +_n \int_{b_1}^b f(x) \times_n \Delta x$$

I5.

Let's

$$f(x) = \frac{dF}{dx}(x)$$

and

$$x, x_i, f(x), F(x), a, b, \Delta x, k, \sum_{i=1}^k {}^n f(x_i) \times_n \Delta x \in W_n$$

We get

$$\int_a^b f(x) \times_n \Delta x = \sum_{i=1}^k {}^n f(x_i) \times_n \Delta x = \sum_{i=1}^k {}^n \frac{dF}{dx}(x_i) \times_n \Delta x =$$

$$= \sum_{i=1}^{k}{}^n((F(x_i +_n \Delta x) -_n F(x_i)) \times_n (\frac{1}{\Delta x})) \times_n \Delta x =$$

$$= \sum_{i=1}^{k}{}^n((F(x_i +_n \Delta x) -_n F(x_i)) \times_n ((\frac{1}{\Delta x}) \times_n \Delta x) +_n \iota_5 =$$

$$= \sum_{i=1}^{k}{}^n((F(x_i +_n \Delta x) -_n F(x_i)) +_n \iota_5 =$$

$$= F(b) -_n F(a) +_n \iota_5$$

where ι_5 is a random variable depends on n and m.
So, the probability of equality

$$\int_a^b \frac{dF}{dx} \times_n \Delta x = F(b) -_n F(a)$$

is less than 1.
So, we proved

Theorem 5.56 *Let's*

$$f(x) = \frac{dF}{dx}(x)$$

and

$$x, x_i, f(x), F(x), a, b, \Delta x, k, \sum_{i=1}^{k}{}^n f(x_i) \times_n \Delta x \in W_n$$

Then

$$\int_a^b \frac{dF}{dx} \times_n \Delta x = F(b) -_n F(a) +_n \iota_5$$

Corollary 5.57 *The probability of equality*

$$\int_a^b \frac{dF}{dx} \times_n \Delta x = F(b) -_n F(a)$$

is less than 1.

I6.
Let's consider a function of two variables $z = f(x, y)$, where $x \in [a, b], y \in [c, d]$ and

$$x, y, z, a, b, c, d \in W_n$$

We call "double integral in Cartesian product $W_n \times W_n$" the following expression

$$\int_c^d \int_a^b (f(x,y) \times_n \Delta x) \times_n \Delta y = \sum_{j=1}^q {}^n(\sum_{i=1}^k {}^n(f(x_i, y_j) \times_n \Delta x)) \times_n \Delta y$$

We assume that segment $[c,d]$ is divided on parts

$$[y_j, y_{j+n}1], j = 1, \ldots, q -_n 1, y_1 = c, y_{q+n1} = d, y_{j+n1} -_n y_j = \Delta y,$$

and

$$q, k, \Delta x, \Delta y, \sum_{i=1}^k {}^n(f(x_i, y_j) \times_n \Delta x), \sum_{j=1}^q {}^n(\sum_{i=1}^k {}^n(f(x_i, y_j) \times_n \Delta x))$$
$$\times_n \Delta y \in W_n$$

We get

$$\sum_{j=1}^q {}^n(\sum_{i=1}^k {}^n(f(x_i, y_j) \times_n \Delta x)) \times_n \Delta y = \sum_{i=1}^k {}^n(\sum_{j=1}^q {}^n(f(x_i, y_j) \times_n \Delta y))$$
$$\times_n \Delta x +_n \iota_6$$

where ι_6 is a random variable depends on n and m.
That means

$$\int_c^d \int_a^b (f(x,y) \times_n \Delta x) \times_n \Delta y = \int_a^b \int_c^d (f(x,y) \times_n \Delta y) \times_n \Delta x +_n \iota_6$$

So, the probability of equality

$$\int_c^d \int_a^b (f(x,y) \times_n \Delta x) \times_n \Delta y = \int_a^b \int_c^d (f(x,y) \times_n \Delta y) \times_n \Delta x$$

is less than 1.
So, we proved

Theorem 5.58 *Let's consider a function of two variables $z = f(x,y)$, where $x \in [a,b], y \in [c,d]$ and*
$$x, y, z, a, b, c, d \in W_n$$
Then

$$\int_c^d \int_a^b (f(x,y) \times_n \Delta x) \times_n \Delta y = \int_a^b \int_c^d (f(x,y) \times_n \Delta y) \times_n \Delta x +_n \iota_6$$

Corollary 5.59 *The probability of equality*

$$\int_c^d \int_a^b (f(x,y) \times_n \Delta x) \times_n \Delta y = \int_a^b \int_c^d (f(x,y) \times_n \Delta y) \times_n \Delta x$$

is less than 1.

I7.

Let's consider a function of three variables $u = f(x, y, z)$, where $x \in [a, b], y \in [c, d], z \in [e, g]$ and

$$x, y, z, u, a, b, c, d, e, g \in W_n$$

We call "triple integral in Cartesian product $W_n \times W_n \times W_n$" the following expression

$$\int_e^g \int_c^d \int_a^b ((f(x,y,z) \times_n \Delta x) \times_n \Delta y) \times_n \Delta z =$$

$$= \sum_{r=1}^p {}^n(\sum_{j=1}^q {}^n(\sum_{i=1}^k {}^n(f(x_i,y_j,z_r) \times_n \Delta x)) \times_n \Delta y) \times_n \Delta z$$

We assume that segment $[e, g]$ is divided on parts

$$[z_r, z_{r+_n 1}], r = 1, \ldots, p -_n 1, z_1 = e, z_{p+_n 1} = g, z_{r+_n 1} -_n z_r = \Delta z,$$

and

$$p, q, k, \Delta x, \Delta y, \Delta z \in W_n$$

$$\sum_{i=1}^k {}^n(f(x_i, y_j, z_r) \times_n \Delta x) \in W_n$$

$$\sum_{j=1}^q {}^n(\sum_{i=1}^k {}^n(f(x_i, y_j, z_r) \times_n \Delta x)) \times_n \Delta y \in W_n$$

$$\sum_{r=1}^p {}^n(\sum_{j=1}^q {}^n(\sum_{i=1}^k {}^n(f(x_i, y_j, z_r) \times_n \Delta x)) \times_n \Delta y) \times_n \Delta z \in W_n$$

We get

$$\sum_{r=1}^p {}^n(\sum_{j=1}^q {}^n(\sum_{i=1}^k {}^n(f(x_i, y_j, z_r) \times_n \Delta x)) \times_n \Delta y) \times_n \Delta z =$$

$$= \sum_{r=1}^p {}^n(\sum_{i=1}^k {}^n(\sum_{j=1}^q {}^n(f(x_i, y_j, z_r) \times_n \Delta y)) \times_n \Delta x) \times_n \Delta z +_n \iota_7$$

where ι_7 is a random variable depends on n and m.

That means

$$\int_e^g \int_c^d \int_a^b ((f(x,y,z) \times_n \Delta x) \times_n \Delta y) \times_n \Delta z =$$

$$= \int_e^g \int_a^b \int_c^d ((f(x,y,z) \times_n \Delta y) \times_n \Delta x) \times_n \Delta z +_n \iota_7$$

We have additional analogous equalities:

$$\int_c^d \int_e^g \int_a^b ((f(x,y,z) \times_n \Delta x) \times_n \Delta z) \times_n \Delta y =$$

$$= \int_c^d \int_a^b \int_e^g ((f(x,y,z) \times_n \Delta z) \times_n \Delta x) \times_n \Delta y +_n \iota_8$$

$$\int_a^b \int_c^d \int_e^g ((f(x,y,z) \times_n \Delta z) \times_n \Delta y) \times_n \Delta x =$$

$$= \int_a^b \int_e^g \int_c^d ((f(x,y,z) \times_n \Delta y) \times_n \Delta z) \times_n \Delta x +_n \iota_9$$

where ι_8, ι_9 are random variables depend on n and m.
So, the probabilities of equalities

$$\int_e^g \int_c^d \int_a^b ((f(x,y,z) \times_n \Delta x) \times_n \Delta y) \times_n \Delta z =$$

$$= \int_e^g \int_a^b \int_c^d ((f(x,y,z) \times_n \Delta y) \times_n \Delta x) \times_n \Delta z$$

$$\int_c^d \int_e^g \int_a^b ((f(x,y,z) \times_n \Delta x) \times_n \Delta z) \times_n \Delta y =$$

$$= \int_c^d \int_a^b \int_e^g ((f(x,y,z) \times_n \Delta z) \times_n \Delta x) \times_n \Delta y$$

$$\int_a^b \int_c^d \int_e^g ((f(x,y,z) \times_n \Delta z) \times_n \Delta y) \times_n \Delta x =$$

$$= \int_a^b \int_e^g \int_c^d ((f(x,y,z) \times_n \Delta y) \times_n \Delta z) \times_n \Delta x$$

are less than 1.

Let's note the values of $\Delta x, \Delta y, \Delta z$ are the integral parameters. The main values of parameters $\Delta x, \Delta y, \Delta z$ are

$$\Delta x_{min} = 0.\underbrace{0...0}_{n-1}2, \Delta y_{min} = 0.\underbrace{0...0}_{n-1}2, \Delta z_{min} = 0.\underbrace{0...0}_{n-1}2$$

So, for given n we have limiting on numbers $a, b, c, d, e, g, q, p, k$.

Let's note triple integral may be defined not only for a function of three variables

$$u = f(x, y, z)$$

but also for vector field of three variables

$$\mathbf{F} = (f_1(x, y, z), f_2(x, y, z), f_3(x, y, z))$$

by analogy with one-dimensional integral.

So, we proved

Theorem 5.60 *Let's consider a function of three variables* $u = f(x, y, z)$, *where* $x \in [a, b], y \in [c, d], z \in [e, g]$ *and*

$$x, y, z, u, a, b, c, d, e, g \in W_n$$

Then

$$\int_e^g \int_c^d \int_a^b ((f(x, y, z) \times_n \Delta x) \times_n \Delta y) \times_n \Delta z =$$

$$= \int_e^g \int_a^b \int_c^d ((f(x, y, z) \times_n \Delta y) \times_n \Delta x) \times_n \Delta z +_n \iota_7$$

$$\int_c^d \int_e^g \int_a^b ((f(x, y, z) \times_n \Delta x) \times_n \Delta z) \times_n \Delta y =$$

$$= \int_c^d \int_a^b \int_e^g ((f(x, y, z) \times_n \Delta z) \times_n \Delta x) \times_n \Delta y +_n \iota_8$$

$$\int_a^b \int_c^d \int_e^g ((f(x, y, z) \times_n \Delta z) \times_n \Delta y) \times_n \Delta x =$$

$$= \int_a^b \int_e^g \int_c^d ((f(x, y, z) \times_n \Delta y) \times_n \Delta z) \times_n \Delta x +_n \iota_9$$

Corollary 5.61 *The probabilities of equalities*

$$\int_e^g \int_c^d \int_a^b ((f(x, y, z) \times_n \Delta x) \times_n \Delta y) \times_n \Delta z =$$

$$= \int_e^g \int_a^b \int_c^d ((f(x, y, z) \times_n \Delta y) \times_n \Delta x) \times_n \Delta z$$

$$\int_c^d \int_e^g \int_a^b ((f(x, y, z) \times_n \Delta x) \times_n \Delta z) \times_n \Delta y =$$

$$= \int_c^d \int_a^b \int_e^g ((f(x, y, z) \times_n \Delta z) \times_n \Delta x) \times_n \Delta y$$

$$\int_a^b \int_c^d \int_e^g ((f(x,y,z) \times_n \Delta z) \times_n \Delta y) \times_n \Delta x =$$

$$= \int_a^b \int_e^g \int_c^d ((f(x,y,z) \times_n \Delta y) \times_n \Delta z) \times_n \Delta x$$

are less than 1.

I8.

Let's consider a function of two variables $z = f_1(x,y) +_n f_2(x,y)$, where $x \in [a,b], y \in [c,d]$ and

$$x, y, z, f_1, f_2, a, b, c, d \in W_n$$

We get

$$\sum_{j=1}^q {}^n(\sum_{i=1}^k {}^n((f_1(x_i,y_j) +_n f_2(x_i,y_j)) \times_n \Delta x)) \times_n \Delta y =$$

$$= \sum_{j=1}^q {}^n(\sum_{i=1}^k {}^n(f_1(x_i,y_j)) \times_n \Delta x)) \times_n \Delta y +_n \sum_{j=1}^q {}^n(\sum_{i=1}^k {}^n(f_2(x_i,y_j) \times_n \Delta x))$$

$$\times_n \Delta y +_n \iota_{10}$$

where ι_{10} is a random variable depends on n and m.
That means

$$\int_c^d \int_a^b ((f_1(x,y) +_n f_2(x,y)) \times_n \Delta x) \times_n \Delta y =$$

$$= \int_c^d \int_a^b (f_1(x,y) \times_n \Delta x) \times_n \Delta y +_n$$

$$+_n \int_c^d \int_a^b (f_2(x,y) \times_n \Delta x) \times_n \Delta y +_n \iota_{10}$$

So, the probability of equality

$$\int_c^d \int_a^b ((f_1(x,y) +_n f_2(x,y)) \times_n \Delta x) \times_n \Delta y =$$

$$= \int_c^d \int_a^b (f_1(x,y) \times_n \Delta x) \times_n \Delta y +_n \int_c^d \int_a^b (f_2(x,y) \times_n \Delta x) \times_n \Delta y$$

is less than 1.
So, we proved

Theorem 5.62 *Let's consider a function of two variables* $z = f_1(x, y) +_n f_2(x, y)$, *where* $x \in [a, b], y \in [c, d]$ *and*

$$x, y, z, f_1, f_2, a, b, c, d \in W_n$$

Then

$$\int_c^d \int_a^b ((f_1(x, y) +_n f_2(x, y)) \times_n \Delta x) \times_n \Delta y =$$

$$= \int_c^d \int_a^b (f_1(x, y) \times_n \Delta x) \times_n \Delta y +_n$$

$$+_n \int_c^d \int_a^b (f_2(x, y) \times_n \Delta x) \times_n \Delta y +_n \iota_{10}$$

Corollary 5.63 *The probability of equality*

$$\int_c^d \int_a^b ((f_1(x, y) +_n f_2(x, y)) \times_n \Delta x) \times_n \Delta y =$$

$$= \int_c^d \int_a^b (f_1(x, y) \times_n \Delta x) \times_n \Delta y +_n \int_c^d \int_a^b (f_2(x, y) \times_n \Delta x) \times_n \Delta y$$

is less than 1.

I9.

Let's consider a function of two variables $z = L \times_n f(x, y)$, where $x \in [a, b], y \in [c, d]$ and

$$x, y, L, z, a, b, c, d \in W_n, L = const$$

We get

$$\sum_{j=1}^q {}^n \left(\sum_{i=1}^k {}^n (L \times_n f(x_i, y_j)) \times_n \Delta x \right) \times_n \Delta y =$$

$$= L \times_n \left(\sum_{j=1}^q {}^n \left(\sum_{i=1}^k {}^n f(x_i, y_j) \right) \times_n \Delta x \right) \times_n \Delta y \right) +_n \iota_{11}$$

where ι_{11} is a random variable depends on n and m.
That means

$$\int_c^d \int_a^b ((L \times_n f(x, y)) \times_n \Delta x) \times_n \Delta y =$$

$$= L \times_n \left(\int_c^d \int_a^b (f(x, y) \times_n \Delta x) \times_n \Delta y \right) +_n \iota_{11}$$

So, the probability of equality

$$\int_c^d \int_a^b ((L \times_n f(x, y)) \times_n \Delta x) \times_n \Delta y =$$

$$= L \times_n \left(\int_c^d \int_a^b (f(x,y) \times_n \Delta x) \times_n \Delta y \right)$$

is less than 1.

So, we proved

Theorem 5.64 *Let's consider a function of two variables* $z = L \times_n f(x,y)$, *where* $x \in [a,b], y \in [c,d]$ *and*

$$x, y, L, z, a, b, c, d \in W_n, L = const$$

Then

$$\int_c^d \int_a^b ((L \times_n f(x,y)) \times_n \Delta x) \times_n \Delta y =$$

$$= L \times_n \left(\int_c^d \int_a^b (f(x,y) \times_n \Delta x) \times_n \Delta y \right) +_n \iota_{11}$$

Corollary 5.65 *The probability of equality*

$$\int_c^d \int_a^b ((L \times_n f(x,y)) \times_n \Delta x) \times_n \Delta y =$$

$$= L \times_n \left(\int_c^d \int_a^b (f(x,y) \times_n \Delta x) \times_n \Delta y \right)$$

is less than 1.

I10.

We have $u = f_1(x,y,z) +_n f_2(x,y,z)$, where $x \in [a,b], y \in [c,d], z \in [e,g]$ and

$$x, y, z, u, f_1 f_2, a, b, c, d, e, g \in W_n$$

let's consider

$$\int_e^g \int_c^d \int_a^b (((f_1(x,y,z) +_n f_2(x,y,z)) \times_n \Delta x) \times_n \Delta y) \times_n \Delta z =$$

$$= \sum_{r=1}^p {}^n \left(\sum_{j=1}^q {}^n \left(\sum_{i=1}^k {}^n ((f_1(x_i,y_j,z_r) +_n f_2(x_i,y_j,z_r)) \times_n \Delta x)) \times_n \Delta y \right) \times_n \Delta z = \right.$$

$$= \sum_{r=1}^p {}^n \left(\sum_{j=1}^q {}^n \left(\sum_{i=1}^k {}^n (f_1(x_i,y_j,z_r) \times_n \Delta x)) \times_n \Delta y \right) \times_n \Delta z +_n \right.$$

$$+_n \sum_{r=1}^p {}^n \left(\sum_{j=1}^q {}^n \left(\sum_{i=1}^k {}^n (f_2(x_i,y_j,z_r) \times_n \Delta x)) \times_n \Delta y \right) \times_n \Delta z +_n \iota_{12} \right.$$

where ι_{12} is a random variable depends on n and m.

$$\int_e^g \int_c^d \int_a^b (((f_1(x,y,z) +_n f_2(x,y,z)) \times_n \Delta x) \times_n \Delta y) \times_n \Delta z =$$

$$= \int_e^g \int_c^d \int_a^b ((f_1(x,y,z) \times_n \Delta x) \times_n \Delta y) \times_n \Delta z +_n$$

$$+_n \int_e^g \int_c^d \int_a^b ((f_2(x,y,z) \times_n \Delta x) \times_n \Delta y) \times_n \Delta z +_n \iota_{12}$$

So, the probability of equality

$$\int_e^g \int_c^d \int_a^b (((f_1(x,y,z) +_n f_2(x,y,z)) \times_n \Delta x) \times_n \Delta y) \times_n \Delta z =$$

$$= \int_e^g \int_c^d \int_a^b ((f_1(x,y,z) \times_n \Delta x) \times_n \Delta y) \times_n \Delta z +_n$$

$$+_n \int_e^g \int_c^d \int_a^b ((f_2(x,y,z) \times_n \Delta x) \times_n \Delta y) \times_n \Delta z$$

is less than 1.

So, we proved

Theorem 5.66 *Let's we have* $u = f_1(x,y,z) +_n f_2(x,y,z)$, *where* $x \in [a,b], y \in [c,d], z \in [e,g]$ *and*

$$x, y, z, u, f_1 f_2, a, b, c, d, e, g \in W_n$$

Then

$$\int_e^g \int_c^d \int_a^b (((f_1(x,y,z) +_n f_2(x,y,z)) \times_n \Delta x) \times_n \Delta y) \times_n \Delta z =$$

$$= \int_e^g \int_c^d \int_a^b ((f_1(x,y,z) \times_n \Delta x) \times_n \Delta y) \times_n \Delta z +_n$$

$$+_n \int_e^g \int_c^d \int_a^b ((f_2(x,y,z) \times_n \Delta x) \times_n \Delta y) \times_n \Delta z +_n \iota_{12}$$

Corollary 5.67 *The probability of equality*

$$\int_e^g \int_c^d \int_a^b (((f_1(x,y,z) +_n f_2(x,y,z)) \times_n \Delta x) \times_n \Delta y) \times_n \Delta z =$$

$$= \int_e^g \int_c^d \int_a^b ((f_1(x,y,z) \times_n \Delta x) \times_n \Delta y) \times_n \Delta z +_n$$

$$+_n \int_e^g \int_c^d \int_a^b ((f_2(x,y,z) \times_n \Delta x) \times_n \Delta y) \times_n \Delta z$$

is less than 1.

I11.

We have $u = L \times_n f(x, y, z)$, where $x \in [a, b], y \in [c, d], z \in [e, g]; L = const$
and

$$x, y, z, L, u, f, a, b, c, d, e, g \in W_n$$

Let's consider

$$\int_e^g \int_c^d \int_a^b ((L \times_n f(x, y, z) \times_n \Delta x) \times_n \Delta y) \times_n \Delta z =$$

$$= \sum_{r=1}^p {}^n (\sum_{j=1}^q {}^n (\sum_{i=1}^k {}^n (L \times_n f(x_i, y_j, z_r) \times_n \Delta x)) \times_n \Delta y) \times_n \Delta z =$$

$$= L \times_n (\sum_{r=1}^p {}^n (\sum_{j=1}^q {}^n (\sum_{i=1}^k {}^n (L \times_n f(x_i, y_j, z_r) \times_n \Delta x)) \times_n \Delta y) \times_n \Delta z) +_n \iota_{13}$$

where ι_{13} is a random variable depends on n and m.
That means

$$\int_e^g \int_c^d \int_a^b ((L \times_n f(x, y, z) \times_n \Delta x) \times_n \Delta y) \times_n \Delta z =$$

$$= L \times_n (\int_e^g \int_c^d \int_a^b ((f(x, y, z) \times_n \Delta x) \times_n \Delta y) \times_n \Delta z) +_n \iota_{13}$$

So, the probability of equality

$$\int_e^g \int_c^d \int_a^b ((L \times_n f(x, y, z) \times_n \Delta x) \times_n \Delta y) \times_n \Delta z =$$

$$= L \times_n (\int_e^g \int_c^d \int_a^b ((f(x, y, z) \times_n \Delta x) \times_n \Delta y) \times_n \Delta z)$$

is less than 1.
So, we proved

Theorem 5.68 *Let's we have* $u = L \times_n f(x, y, z)$, *where* $x \in [a, b], y \in [c, d], z \in [e, g]; L = const$ *and*

$$x, y, z, L, u, f, a, b, c, d, e, g \in W_n$$

Then

$$\int_e^g \int_c^d \int_a^b ((L \times_n f(x, y, z) \times_n \Delta x) \times_n \Delta y) \times_n \Delta z =$$

$$= L \times_n (\int_e^g \int_c^d \int_a^b ((f(x, y, z) \times_n \Delta x) \times_n \Delta y) \times_n \Delta z) +_n \iota_{13}$$

Corollary 5.69 *The probability of equality*

$$\int_e^g \int_c^d \int_a^b ((L \times_n f(x,y,z) \times_n \Delta x) \times_n \Delta y) \times_n \Delta z =$$

$$= L \times_n \left(\int_e^g \int_c^d \int_a^b ((f(x,y,z) \times_n \Delta x) \times_n \Delta y) \times_n \Delta z \right)$$

is less than 1.

I12

We have function $f(x,y,z)$, where $x \in [a,b], y \in [c,d], z \in [e,g]$, $\partial f/\partial x$ exists and $\in W_n$ for any point (x,y,z) inside cube $[a,b] \times [c,d] \times [e,g]$, and

$$x, y, z, f, a, b, c, d, e, g \in W_n$$

By **I5** and **I10** we get

$$\int_e^g \int_c^d \int_a^b ((\partial f/\partial x(x,y,z) \times_n \Delta x) \times_n \Delta y) \times_n \Delta z =$$

$$= \int_e^g \int_c^d (f(b,y,z) -_n f(a,y,z)) \times_n \Delta y) \times_n \Delta z +_n \iota_{14} =$$

$$= \int_e^g \int_c^d (f(b,y,z) \times_n \Delta y) \times_n \Delta z -_n$$

$$-_n \int_e^g \int_c^d (f(a,y,z) \times_n \Delta y) \times_n \Delta z +_n \iota_{14} +_n \iota_{15}$$

where ι_{14}, ι_{15} are random variables depend on n and m.
So, the probability of equality

$$\int_e^g \int_c^d \int_a^b ((\partial f/\partial x(x,y,z) \times_n \Delta x) \times_n \Delta y) \times_n \Delta z =$$

$$= \int_e^g \int_c^d (f(b,y,z) \times_n \Delta y) \times_n \Delta z -_n \int_e^g \int_c^d (f(a,y,z) \times_n \Delta y) \times_n \Delta z$$

is less than 1.
So, we proved

Theorem 5.70 *Let's we have function $f(x,y,z)$, where $x \in [a,b], y \in [c,d], z \in [e,g]$, $\partial f/\partial x$ exists and $\in W_n$ for any point (x,y,z) inside cube $[a,b] \times [c,d] \times [e,g]$, and*

$$x, y, z, f, a, b, c, d, e, g \in W_n$$

Then

$$\int_e^g \int_c^d \int_a^b ((\partial f/\partial x(x,y,z) \times_n \Delta x) \times_n \Delta y) \times_n \Delta z =$$

$$= \int_e^g \int_c^d (f(b,y,z) -_n f(a,y,z)) \times_n \Delta y) \times_n \Delta z +_n \iota_{14} =$$

$$= \int_e^g \int_c^d (f(b,y,z) \times_n \Delta y) \times_n \Delta z -_n$$

$$-_n \int_e^g \int_c^d (f(a,y,z) \times_n \Delta y) \times_n \Delta z +_n \iota_{14} +_n \iota_{15}$$

Corollary 5.71 *The probability of equality*

$$\int_e^g \int_c^d \int_a^b ((\partial f/\partial x(x,y,z) \times_n \Delta x) \times_n \Delta y) \times_n \Delta z =$$

$$= \int_e^g \int_c^d (f(b,y,z) \times_n \Delta y) \times_n \Delta z -_n \int_e^g \int_c^d (f(a,y,z) \times_n \Delta y) \times_n \Delta z$$

is less than 1.

I13

Following **D9** we have vector field

$$\mathbf{F} = (f_1(x,y,z), f_2(x,y,z), f_3(x,y,z))$$

with f_1, f_2, f_3, x, y, z belong to W_n. We assume that f_1, f_2, f_3 are differentiable functions, and all partial derivatives belong to W_n.

We consider now divergence, which is a scalar function $\mathbf{F}$

$$div\mathbf{F} = (\nabla, \mathbf{F}) = \partial f_1/\partial x +_n \partial f_2/\partial y +_n \partial f_3/\partial z$$

Let's assume
$x \in [a,b], y \in [c,d], z \in [e,g]$, and

$$x,y,z,f_1,f_2,f_3,a,b,c,d,e,g \in W_n$$

By **I5**, **I7** and **I10** we get

$$\int_e^g \int_c^d \int_a^b ((div\mathbf{F} \times_n \Delta x) \times_n \Delta y) \times_n \Delta z =$$

$$= \int_e^g \int_c^d \int_a^b (((\partial f_1/\partial x +_n \partial f_2/\partial y +_n \partial f_3/\partial z) \times_n \Delta x) \times_n \Delta y) \times_n \Delta z =$$

$$= \int_e^g \int_c^d \int_a^b ((\partial f_1/\partial x \times_n \Delta x) \times_n \Delta y) \times_n \Delta z +_n$$

$$+_n \int_e^g \int_c^d \int_a^b ((\partial f_2/\partial y \times_n \Delta x) \times_n \Delta y) \times_n \Delta z +_n$$

$$+_n \int_e^g \int_c^d \int_a^b ((\partial f_3/\partial z \times_n \Delta x) \times_n \Delta y) \times_n \Delta z +_n \iota_{16} =$$

$$= \int_e^g \int_c^d \int_a^b ((\partial f_1/\partial x \times_n \Delta x) \times_n \Delta y) \times_n \Delta z +_n$$

$$+_n \int_e^g \int_a^b \int_c^d ((\partial f_2/\partial y \times_n \Delta y) \times_n \Delta x) \times_n \Delta z +_n \iota_{17} +_n$$

$$+_n \int_a^b \int_c^d \int_e^g ((\partial f_3/\partial z \times_n \Delta z) \times_n \Delta y) \times_n \Delta x +_n \iota_{18} +_n \iota_{18} =$$

$$= \int_e^g \int_c^d ((f_1(b,y,z) -_n f_1(a,y,z)) \times_n \Delta y) \times_n \Delta z +_n \iota_{19} +_n$$

$$+_n \int_e^g \int_a^b ((f_2(x,d,z) -_n f_2(x,c,z)) \times_n \Delta x) \times_n \Delta z +_n \iota_{20} +_n \iota_{17} +_n$$

$$+_n \int_a^b \int_c^d ((f_3(x,y,g) -_n f_3(x,y,e)) \times_n \Delta y) \times_n \Delta x +_n \iota_{21} +_n \iota_{18} +_n \iota_{16} =$$

$$= \int_e^g \int_c^d (f_1(b,y,z) \times_n \Delta y) \times_n \Delta z -_n$$

$$-_n \int_e^g \int_c^d (f_1(a,y,z) \times_n \Delta y) \times_n \Delta z +_n \iota_{22} +_n \iota_{19} +_n$$

$$+_n \int_e^g \int_a^b (f_2(x,d,z) \times_n \Delta x) \times_n \Delta z -_n$$

$$-_n \int_e^g \int_a^b (f_2(x,c,z) \times_n \Delta x) \times_n \Delta z +_n \iota_{20} +_n \iota_{23} +_n \iota_{17} +_n$$

$$+_n \int_a^b \int_c^d (f_3(x,y,g) \times_n \Delta y) \times_n \Delta x -_n$$

$$-_n \int_a^b \int_c^d (f_3(x,y,e) \times_n \Delta y) \times_n \Delta x +_n \iota_{24} +_n \iota_{21} +_n \iota_{18} +_n \iota_{16}$$

where $\iota_{16}, \iota_{17}, \iota_{18}, \iota_{19}, \iota_{20}, \iota_{21}, \iota_{22}, \iota_{23}, \iota_{24}$ are random variables depend on n and m.

So, the probability of equality

$$\int_e^g \int_c^d \int_a^b ((div\mathbf{F} \times_n \Delta x) \times_n \Delta y) \times_n \Delta z =$$

$$= \int_e^g \int_c^d (f_1(b,y,z) \times_n \Delta y) \times_n \Delta z -_n \int_e^g \int_c^d (f_1(a,y,z) \times_n \Delta y) \times_n \Delta z +_n$$

$$+_n \int_e^g \int_a^b (f_2(x,d,z) \times_n \Delta x) \times_n \Delta z -_n \int_e^g \int_a^b (f_2(x,c,z) \times_n \Delta x) \times_n \Delta z +_n$$

$$+_n \int_a^b \int_c^d (f_3(x,y,g) \times_n \Delta y) \times_n \Delta x -_n \int_a^b \int_c^d (f_3(x,y,e) \times_n \Delta y) \times_n \Delta x$$

is less than 1.

So, we proved

Theorem 5.72 *Let's we have vector field*

$$\mathbf{F} = (f_1(x,y,z), f_2(x,y,z), f_3(x,y,z))$$

with f_1, f_2, f_3, x, y, z belong to W_n. We assume that f_1, f_2, f_3 are differentiable functions, and all partial derivatives belong to W_n. Then

$$\int_e^g \int_c^d \int_a^b ((div\mathbf{F} \times_n \Delta x) \times_n \Delta y) \times_n \Delta z =$$

$$= \int_e^g \int_c^d (f_1(b,y,z) \times_n \Delta y) \times_n \Delta z -_n$$

$$-_n \int_e^g \int_c^d (f_1(a,y,z) \times_n \Delta y) \times_n \Delta z +_n \iota_{22} +_n \iota_{19} +_n$$

$$+_n \int_e^g \int_a^b (f_2(x,d,z) \times_n \Delta x) \times_n \Delta z -_n$$

$$-_n \int_e^g \int_a^b (f_2(x,c,z) \times_n \Delta x) \times_n \Delta z +_n \iota_{20} +_n \iota_{23} +_n \iota_{17} +_n$$

$$+_n \int_a^b \int_c^d (f_3(x,y,g) \times_n \Delta y) \times_n \Delta x -_n$$

$$-_n \int_a^b \int_c^d (f_3(x,y,e) \times_n \Delta y) \times_n \Delta x +_n \iota_{24} +_n \iota_{21} +_n \iota_{18} +_n \iota_{16}$$

Corollary 5.73 *The probability of equality*

$$\int_e^g \int_c^d \int_a^b ((div\mathbf{F} \times_n \Delta x) \times_n \Delta y) \times_n \Delta z =$$

$$= \int_e^g \int_c^d (f_1(b,y,z) \times_n \Delta y) \times_n \Delta z -_n \int_e^g \int_c^d (f_1(a,y,z) \times_n \Delta y) \times_n \Delta z +_n$$

$$+_n \int_e^g \int_a^b (f_2(x,d,z) \times_n \Delta x) \times_n \Delta z -_n \int_e^g \int_a^b (f_2(x,c,z) \times_n \Delta x) \times_n \Delta z +_n$$

$$+_n \int_a^b \int_c^d (f_3(x,y,g) \times_n \Delta y) \times_n \Delta x -_n \int_a^b \int_c^d (f_3(x,y,e) \times_n \Delta y) \times_n \Delta x$$

is less than 1.

I14

We have function $f(x,y)$, where $y \in [a,b]$, $\partial f/\partial x$ exists and $\in W_n$ for any point (x,y) inside $W_n \times [a,b]$, and

$$x, y, f, a, b \in W_n$$

By **D1** and **D3** we get

$$\frac{d}{dx}\left(\int_a^b f(x,y) \times_n \Delta y\right) = \frac{d}{dx}\left(\sum_{i=1}^k {}^n f(x,y_i) \times_n \Delta y\right) =$$

$$= \sum_{i=1}^k {}^n \partial f(x,y_i)/\partial x \times_n \Delta y +_n \iota_{25} =$$

$$= \int_a^b \partial f(x,y)/\partial x \times_n \Delta y +_n \iota_{25}$$

where ι_{25} is a random variable depends on n and m.
That means

$$\frac{d}{dx}\left(\int_a^b f(x,y) \times_n \Delta y\right) = \int_a^b \partial f(x,y)/\partial x \times_n \Delta y +_n \iota_{25}$$

So, the probability of equality

$$\frac{d}{dx}\left(\int_a^b f(x,y) \times_n \Delta y\right) = \int_a^b \partial f(x,y)/\partial x \times_n \Delta y$$

is less than 1.
So, we proved

Theorem 5.74 *Let's we have function $f(x, y)$, where $y \in [a, b]$, $\partial f / \partial x$ exists and $\in W_n$ for any point (x, y) inside $W_n \times [a, b]$, and*

$$x, y, f, a, b \in W_n$$

Then

$$\frac{d}{dx}\left(\int_a^b f(x, y) \times_n \Delta y\right) = \int_a^b \partial f(x, y)/\partial x \times_n \Delta y +_n \iota_{25}$$

Corollary 5.75 *The probability of equality*

$$\frac{d}{dx}\left(\int_a^b f(x, y) \times_n \Delta y\right) = \int_a^b \partial f(x, y)/\partial x \times_n \Delta y$$

is less than 1.

I15

We have function $f(x, y, z)$, where $x \in [a, b], y \in [c, d], z \in [e, g]$, $\partial f / \partial x, \partial f / \partial y, \partial f / \partial z$ exist and $\in W_n$ for any point (x, y, z) inside cube $[a, b] \times [c, d] \times [e, g]$, and

$$x, y, z, f, a, b, c, d, e, g \in W_n$$

By **I12** we have

$$\int_e^g \int_c^d \int_a^b ((\mathbf{grad}f(x, y, z) \times_n \Delta x) \times_n \Delta y) \times_n \Delta z =$$

$$= \left(\int_e^g \int_c^d (f(b, y, z) \times_n \Delta y) \times_n \Delta z -_n \int_e^g \int_c^d (f(a, y, z) \times_n \Delta y) \times_n \Delta z\right),$$

$$\int_e^g \int_a^b (f(x, d, z) \times_n \Delta x) \times_n \Delta z -_n \int_e^g \int_a^b (f(x, c, z) \times_n \Delta x) \times_n \Delta z,$$

$$\int_a^b \int_c^d (f(x, y, g) \times_n \Delta y) \times_n \Delta x -_n \int_a^b \int_c^d (f(x, y, e) \times_n \Delta y) \times_n \Delta x) +_n$$

$$+_n (\iota_{26}, \iota_{27}, \iota_{28})$$

where $(\iota_{26}, \iota_{27}, \iota_{28})$ is a random vector $\in E_3 W_n$ and $\iota_{26}, \iota_{27}, \iota_{28}$ are random variables depend on n and m.

So, we proved

Theorem 5.76 *Let's we have function $f(x, y, z)$, where $x \in [a, b], y \in [c, d], z \in [e, g]$, $\partial f/\partial x, \partial f/\partial y, \partial f/\partial z$ exist and $\in W_n$ for any point (x, y, z) inside cube $[a, b] \times [c, d] \times [e, g]$, and*

$$x, y, z, f, a, b, c, d, e, g \in W_n$$

Then

$$\int_e^g \int_c^d \int_a^b ((\boldsymbol{grad}f(x, y, z) \times_n \Delta x) \times_n \Delta y) \times_n \Delta z =$$

$$= (\int_e^g \int_c^d (f(b, y, z) \times_n \Delta y) \times_n \Delta z -_n \int_e^g \int_c^d (f(a, y, z) \times_n \Delta y) \times_n \Delta z),$$

$$\int_e^g \int_a^b (f(x, d, z) \times_n \Delta x) \times_n \Delta z -_n \int_e^g \int_a^b (f(x, c, z) \times_n \Delta x) \times_n \Delta z,$$

$$\int_a^b \int_c^d (f(x, y, g) \times_n \Delta y) \times_n \Delta x -_n \int_a^b \int_c^d (f(x, y, e) \times_n \Delta y) \times_n \Delta x) +_n$$

$$+_n(\iota_{26}, \iota_{27}, \iota_{28})$$

Corollary 5.77 *The probability of equality*

$$\int_e^g \int_c^d \int_a^b ((\boldsymbol{grad}f(x, y, z) \times_n \Delta x) \times_n \Delta y) \times_n \Delta z =$$

$$= (\int_e^g \int_c^d (f(b, y, z) \times_n \Delta y) \times_n \Delta z -_n \int_e^g \int_c^d (f(a, y, z) \times_n \Delta y) \times_n \Delta z),$$

$$\int_e^g \int_a^b (f(x, d, z) \times_n \Delta x) \times_n \Delta z -_n \int_e^g \int_a^b (f(x, c, z) \times_n \Delta x) \times_n \Delta z,$$

$$\int_a^b \int_c^d (f(x, y, g) \times_n \Delta y) \times_n \Delta x -_n \int_a^b \int_c^d (f(x, y, e) \times_n \Delta y) \times_n \Delta x)$$

is less than 1.

6

Observability and Thermodynamical Equations

Classic Thermodynamical equations are well presented in (see [10]). Now we consider them from Mathematics with Observers point of view.

We consider thermal natural variables: T – temperature and S – entropy, and mechanical natural variables: P – pressure and V – volume.

Also we consider thermodynamic potentials as functions of their thermal and mechanical variables: the internal energy $U = U(S, V)$ with given T and P considered as the parameters, enthalpy $H = H(S, P)$ with given T and V considered as the parameters, Helmholtz free energy $A = A(T, V)$ with given S and P considered as the parameters, and Gibbs free energy $G = G(T, P)$ with given S and V considered as the parameters.

The combined first and second thermodynamic laws give equation written in Mathematics with Observers arithmetic:

(TD1)

$$\Delta U = T \times_n \Delta S -_n P \times_n \Delta V$$

Definition of enthalpy can be written as

(TD2)

$$H = U +_n P \times_n V$$

Definition of Helmholtz free energy can be written as

(TD3)

$$A = U +_n T \times_n S$$

Definition of Gibbs free energy can be written as

(TD4)

$$G = H +_n T \times_n S$$

(TD2') We get

$$\Delta H = \Delta U +_n P \times_n \Delta V +_n V \times_n \Delta P +_n \eta_1 =$$

DOI: 10.1201/9781003175902-6

$$= T \times_n \Delta S -_n P \times_n \Delta V +_n P \times_n \Delta V +_n V \times_n \Delta P +_n \eta_1 =$$

$$= T \times_n \Delta S +_n V \times_n \Delta P +_n \eta_1$$

I.e.

$$\Delta H = T \times_n \Delta S +_n V \times_n \Delta P +_n \eta_1$$

where η_1 is a random variable depends on n and m.
We assume that all elements of this equality belong to W_n.
So, the probability of equality

$$\Delta H = T \times_n \Delta S +_n V \times_n \Delta P$$

is less than 1.
So, we proved

Theorem 6.1

$$\Delta H = T \times_n \Delta S +_n V \times_n \Delta P +_n \eta_1$$

We assume that all elements of this equality belong to W_n.

Corollary 6.2 *The probability of equality*

$$\Delta H = T \times_n \Delta S +_n V \times_n \Delta P$$

is less than 1.

(**TD3'**)
We get

$$\Delta A = \Delta U -_n T \times_n \Delta S -_n S \times_n \Delta T +_n \eta_2 =$$

$$= T \times_n \Delta S -_n P \times_n \Delta V -_n T \times_n \Delta S -_n S \times_n \Delta T +_n \eta_2 =$$

$$= -P \times_n \Delta V -_n S \times_n \Delta T +_n \eta_2$$

I.e.

$$\Delta A = -P \times_n \Delta V -_n S \times_n \Delta T +_n \eta_2$$

where η_2 is a random variable depends on n and m.
We assume that all elements of this equality belong to W_n.
So, the probability of equality

$$\Delta A = -S \times_n \Delta T -_n P \times_n \Delta V$$

is less than 1.
So, we proved

Theorem 6.3

$$\Delta A = -P \times_n \Delta V -_n S \times_n \Delta T +_n \eta_2$$

We assume that all elements of this equality belong to W_n.

Corollary 6.4 *The probability of equality*

$$\Delta A = -S \times_n \Delta T -_n P \times_n \Delta V$$

is less than 1.

(**TD4'**)
We get

$$\Delta G = \Delta H -_n T \times_n \Delta S -_n S \times_n \Delta T +_n \eta_3 =$$

$$= T \times_n \Delta S +_n V \times_n \Delta P +_n \eta_1 -_n T \times_n \Delta S -_n S \times_n \Delta T +_n \eta_3 =$$

$$= V \times_n \Delta P +_n \eta_1 -_n S \times_n \Delta T +_n \eta_3$$

I.e.

$$\Delta G = V \times_n \Delta P -_n S \times_n \Delta T +_n \eta_4$$

where η_3, η_4 are the random variables depend on n and m.
We assume that all elements of this equality belong to W_n.
So, the probability of equality

$$\Delta G = -S \times_n \Delta T +_n V \times_n \Delta P$$

is less than 1.
So, we proved

Theorem 6.5

$$\Delta G = V \times_n \Delta P -_n S \times_n \Delta T +_n \eta_4$$

We assume that all elements of this equality belong to W_n.

Corollary 6.6 *The probability of equality*

$$\Delta G = -S \times_n \Delta T +_n V \times_n \Delta P$$

is less than 1.

(TD1')
By (**TD1**) we get

$$T = \partial U/\partial S +_n \eta_5$$

and

$$-P = \partial U/\partial V +_n \eta_6$$

where η_5, η_6 are random variables depend on n and m.
So, the probabilities of equalities

$$T = \partial U/\partial S_5$$

and

$$-P = \partial U/\partial V$$

are less than 1.
So, we proved

Theorem 6.7
$$T = \partial U/\partial S +_n \eta_5$$

and

$$-P = \partial U/\partial V +_n \eta_6$$

We assume that all elements of these equalities belong to W_n.

Corollary 6.8 *The probabilities of equalities*

$$T = \partial U/\partial S_5$$

and

$$-P = \partial U/\partial V$$

are less than 1.

(TD5)
And we get

$$\partial T/\partial V = -\partial P/\partial S +_n \eta_7$$

where η_7 is a random variable depends on n and m.
We assume that all elements of (**TD5**) belong to W_n.
So, the probability of equality

$$\partial T/\partial V = -\partial P/\partial S$$

is less than 1.
So, we proved

Theorem 6.9

$$\partial T/\partial V = -\partial P/\partial S +_n \eta_7$$

We assume that all elements of this equality belong to W_n.

Corollary 6.10 *The probability of equality*

$$\partial T/\partial V = -\partial P/\partial S$$

is less than 1.

(TD2")
By **(TD2')** we get

$$T = \partial H/\partial S +_n \eta_8$$

and

$$V = \partial H/\partial P +_n \eta_9$$

where η_8, η_9 are random variables depend on n and m.
So, the probabilities of equalities

$$T = \partial H/\partial S$$

and

$$V = \partial H/\partial P$$

are less than 1.
So, we proved

Theorem 6.11

$$T = \partial H/\partial S +_n \eta_8$$

and

$$V = \partial H/\partial P +_n \eta_9$$

We assume that all elements of these equalities belong to W_n.

Corollary 6.12 *The probabilities of equalities*

$$T = \partial H/\partial S$$

and

$$V = \partial H/\partial P$$

are less than 1.

(TD6)
And we get

$$\partial T/\partial P = \partial V/\partial S +_n \eta_{10}$$

where η_{10} is a random variable depends on n and m.
We assume that all elements of **(TD6)** belong to W_n.
So, the probability of equality

$$\partial T/\partial P = \partial V/\partial S$$

is less than 1.
So, we proved

Theorem 6.13

$$\partial T/\partial P = \partial V/\partial S +_n \eta_{10}$$

We assume that all elements of this equality belong to W_n.

Corollary 6.14 *The probability of equality*

$$\partial T/\partial P = \partial V/\partial S$$

is less than 1.

(TD3")
By **(TD3')** we get

$$-S = \partial A/\partial T +_n \eta_{11}$$

and

$$-P = \partial A/\partial V +_n \eta_{12}$$

where η_{11}, η_{12} are random variables depend on n and m.
So, the probabilities of equalities

$$-S = \partial A/\partial T$$

and

$$-P = \partial A/\partial V$$

are less than 1.
So, we proved

Theorem 6.15

$$-S = \partial A/\partial T +_n \eta_{11}$$

and

$$-P = \partial A/\partial V +_n \eta_{12}$$

We assume that all elements of these equalities belong to W_n.

Corollary 6.16 *The probabilities of equalities*

$$-S = \partial A/\partial T$$

and

$$-P = \partial A/\partial V$$

are less than 1.

(TD7)
And we get

$$\partial S/\partial V = \partial P/\partial T +_n \eta_{13}$$

where η_{13} is a random variable depends on n and m.
We assume that all elements of **(TD7)** belong to W_n.
So, the probability of equality

$$\partial S/\partial V = \partial P/\partial T$$

is less than 1.
So, we proved

Theorem 6.17

$$\partial S/\partial V = \partial P/\partial T +_n \eta_{13}$$

We assume that all elements of this equality belong to W_n.

Corollary 6.18 *The probability of equality*

$$\partial S/\partial V = \partial P/\partial T$$

is less than 1.

(TD4'')
By **(TD4')** we get

$$-S = \partial G/\partial T +_n \eta_{14}$$

and

$$V = \partial G/\partial P +_n \eta_{15}$$

where η_{14}, η_{15} are random variables depend on n and m.
So, the probabilities of equalities

$$-S = \partial G/\partial T$$

and

$$V = \partial G/\partial P$$

are less than 1.
So, we proved

Theorem 6.19

$$-S = \partial G/\partial T +_n \eta_{14}$$

and

$$V = \partial G/\partial P +_n \eta_{15}$$

We assume that all elements of these equalities belong to W_n.

Corollary 6.20 *The probabilities of equalities*

$$-S = \partial G/\partial T$$

and

$$V = \partial G/\partial P$$

are less than 1.

(**TD8**)
And we get

$$-\partial S/\partial P = \partial V/\partial T +_n \eta_{16}$$

where η_{16} is a random variable depends on n and m.
We assume that all elements of (**TD8**) belong to W_n.
So, the probability of equality

$$-\partial S/\partial P = \partial V/\partial T$$

is less than 1.
So, we proved

Theorem 6.21

$$-\partial S/\partial P = \partial V/\partial T +_n \eta_{16}$$

We assume that all elements of this equality belong to W_n.

Corollary 6.22 *The probability of equality*

$$-\partial S/\partial P = \partial V/\partial T$$

is less than 1.

7

Observability and Equation of Continuity

The equations are derived in Mathematics with Observers using classic way, i.e. from the basic principles of continuity of mass, momentum, and energy. For that matter, sometimes it is necessary to consider a finite arbitrary volume, called a control volume, over which these principles can be applied. The control volume can remain fixed in space or can move with the fluid.

Let's now we have the ideal fluid velocity

$$\mathbf{v}(x, y, z, t), \mathbf{v} \in E_3 W_n, x, y, z, t \in W_n$$

and two thermodynamic quantities - pressure

$$p(x, y, z, t), p, x, y, z, t \in W_n$$

and density

$$\rho(x, y, z, t), \rho, x, y, z, t \in W_n$$

Let's consider the volume

$$[a, b] \times [c, d] \times [e, g] \in W_n \times W_n \times W_n, x \in [a, b], y \in [c, d], z \in [e, g]$$

and

$$x, y, z, a, b, c, d, e, g \in W_n, a < b, c < d, e < g$$

The mass of fluid in this volume is

$$\int_e^g \int_c^d \int_a^b (\rho \times_n \Delta x) \times_n \Delta y) \times_n \Delta z$$

The total mass of fluid flowing in and out of this volume in unit time is

$$\int_e^g \int_c^d (\rho \times_n v_1(b, y, z) \times_n \Delta y) \times_n \Delta z -_n$$

$$-_n \int_e^g \int_c^d (\rho \times_n v_1(a, y, z) \times_n \Delta y) \times_n \Delta z +_n$$

DOI: 10.1201/9781003175902-7

$$+_n \int_e^g \int_a^b (\rho \times_n v_2(x,d,z) \times_n \Delta x) \times_n \Delta z -_n$$

$$-_n \int_e^g \int_a^b (\rho \times_n v_2(x,c,z) \times_n \Delta x) \times_n \Delta z +_n$$

$$+_n \int_a^b \int_c^d (\rho \times_n v_3(x,y,g) \times_n \Delta y) \times_n \Delta x -_n$$

$$-_n \int_a^b \int_c^d (\rho \times_n v_3(x,y,e) \times_n \Delta y) \times_n \Delta x$$

where $\mathbf{v} = (v_1, v_2, v_3)$ and $v_1, v_2, v_3 \in W_n$.
Decrease per unit time in the mass of fluid in this volume is

$$-\partial/\partial t \int_e^g \int_c^d \int_a^b (\rho \times_n \Delta x) \times_n \Delta y) \times_n \Delta z$$

So, we have

$$-\partial/\partial t \int_e^g \int_c^d \int_a^b (\rho \times_n \Delta x) \times_n \Delta y) \times_n \Delta z =$$

$$= \int_e^g \int_c^d (\rho \times_n v_1(b,y,z) \times_n \Delta y) \times_n \Delta z -_n$$

$$-_n \int_e^g \int_c^d (\rho \times_n v_1(a,y,z) \times_n \Delta y) \times_n \Delta z +_n$$

$$+_n \int_e^g \int_a^b (\rho \times_n v_2(x,d,z) \times_n \Delta x) \times_n \Delta z -_n$$

$$-_n \int_e^g \int_a^b (\rho \times_n v_2(x,c,z) \times_n \Delta x) \times_n \Delta z +_n$$

$$+_n \int_a^b \int_c^d (\rho \times_n v_3(x,y,g) \times_n \Delta y) \times_n \Delta x -_n$$

$$-_n \int_a^b \int_c^d (\rho \times_n v_3(x,y,e) \times_n \Delta y) \times_n \Delta x$$

By **I13** we get

$$\int_e^g \int_c^d (\rho \times_n v_1(b,y,z) \times_n \Delta y) \times_n \Delta z -_n$$

$$-_n \int_e^g \int_c^d (\rho \times_n v_1(a,y,z) \times_n \Delta y) \times_n \Delta z +_n$$

$$+_n \int_e^g \int_a^b (\rho \times_n v_2(x,d,z) \times_n \Delta x) \times_n \Delta z -_n$$

$$-_n \int_e^g \int_a^b (\rho \times_n v_2(x,c,z) \times_n \Delta x) \times_n \Delta z +_n$$

$$+_n \int_a^b \int_c^d (\rho \times_n v_3(x,y,g) \times_n \Delta y) \times_n \Delta x -_n$$

$$-_n \int_a^b \int_c^d (\rho \times_n v_3(x,y,e) \times_n \Delta y) \times_n \Delta x =$$

$$= \int_e^g \int_c^d \int_a^b ((div(\rho \times_n \mathbf{v}) \times_n \Delta x) \times_n \Delta y) \times_n \Delta z +_n \omega_1$$

where ω_1 is a random variable depends on n and m.
So, the probability of equation

$$\int_e^g \int_c^d (\rho \times_n v_1(b,y,z) \times_n \Delta y) \times_n \Delta z -_n$$

$$-_n \int_e^g \int_c^d (\rho \times_n v_1(a,y,z) \times_n \Delta y) \times_n \Delta z +_n$$

$$+_n \int_e^g \int_a^b (\rho \times_n v_2(x,d,z) \times_n \Delta x) \times_n \Delta z -_n$$

$$-_n \int_e^g \int_a^b (\rho \times_n v_2(x,c,z) \times_n \Delta x) \times_n \Delta z +_n$$

$$+_n \int_a^b \int_c^d (\rho \times_n v_3(x,y,g) \times_n \Delta y) \times_n \Delta x -_n$$

$$-_n \int_a^b \int_c^d (\rho \times_n v_3(x,y,e) \times_n \Delta y) \times_n \Delta x =$$

$$= \int_e^g \int_c^d \int_a^b ((div(\rho \times_n \mathbf{v}) \times_n \Delta x) \times_n \Delta y) \times_n \Delta z$$

is less than 1.
So, we proved

Theorem 7.1

$$\int_e^g \int_c^d (\rho \times_n v_1(b, y, z) \times_n \Delta y) \times_n \Delta z -_n$$

$$-_n \int_e^g \int_c^d (\rho \times_n v_1(a, y, z) \times_n \Delta y) \times_n \Delta z +_n$$

$$+_n \int_e^g \int_a^b (\rho \times_n v_2(x, d, z) \times_n \Delta x) \times_n \Delta z -_n$$

$$-_n \int_e^g \int_a^b (\rho \times_n v_2(x, c, z) \times_n \Delta x) \times_n \Delta z +_n$$

$$+_n \int_a^b \int_c^d (\rho \times_n v_3(x, y, g) \times_n \Delta y) \times_n \Delta x -_n$$

$$-_n \int_a^b \int_c^d (\rho \times_n v_3(x, y, e) \times_n \Delta y) \times_n \Delta x =$$

$$= \int_e^g \int_c^d \int_a^b ((div(\rho \times_n v) \times_n \Delta x) \times_n \Delta y) \times_n \Delta z +_n \omega_1$$

We assume that all elements of this equality belong to W_n.

Corollary 7.2 *The probability of equality*

$$\int_e^g \int_c^d (\rho \times_n v_1(b, y, z) \times_n \Delta y) \times_n \Delta z -_n$$

$$-_n \int_e^g \int_c^d (\rho \times_n v_1(a, y, z) \times_n \Delta y) \times_n \Delta z +_n$$

$$+_n \int_e^g \int_a^b (\rho \times_n v_2(x, d, z) \times_n \Delta x) \times_n \Delta z -_n$$

$$-_n \int_e^g \int_a^b (\rho \times_n v_2(x, c, z) \times_n \Delta x) \times_n \Delta z +_n$$

$$+_n \int_a^b \int_c^d (\rho \times_n v_3(x, y, g) \times_n \Delta y) \times_n \Delta x -_n$$

$$-_n \int_a^b \int_c^d (\rho \times_n v_3(x, y, e) \times_n \Delta y) \times_n \Delta x =$$

$$= \int_e^g \int_c^d \int_a^b ((div(\rho \times_n v) \times_n \Delta x) \times_n \Delta y) \times_n \Delta z$$

is less than 1.

By **I14** and **I10** we get

$$\int_e^g \int_c^d \int_a^b \left(\left(\left(\partial\rho/\partial t +_n \ div(\rho \times_n \mathbf{v})\right) \times_n \Delta x\right) \times_n \Delta y\right) \times_n \Delta z = \omega_1 +_n \omega_2$$

where ω_2 is a random variable depends on n and m.
So, the probability of equation

$$\int_e^g \int_c^d \int_a^b \left(\left(\left(\partial\rho/\partial t +_n \ div(\rho \times_n \mathbf{v})\right) \times_n \Delta x\right) \times_n \Delta y\right) \times_n \Delta z = 0$$

is less than 1.
So, we proved

Theorem 7.3

$$\int_e^g \int_c^d \int_a^b \left(\left(\left(\partial\rho/\partial t +_n \ div(\rho \times_n v)\right) \times_n \Delta x\right) \times_n \Delta y\right) \times_n \Delta z = \omega_1 +_n \omega_2$$

We assume that all elements of this equality belong to W_n.

Corollary 7.4 *The probability of equality*

$$\int_e^g \int_c^d \int_a^b \left(\left(\left(\partial\rho/\partial t +_n \ div(\rho \times_n v)\right) \times_n \Delta x\right) \times_n \Delta y\right) \times_n \Delta z = 0$$

is less than 1.

Since this equation must hold for any volume (for any possible $a, b, c, d, e, g \in W_n$) we get

$$\partial\rho/\partial t +_n \ div(\rho \times_n \mathbf{v}) = \omega_3$$

where ω_3 is a random variable depends on n and m.
So, the probability of equation

$$\partial\rho/\partial t +_n \ div(\rho \times_n \mathbf{v}) = 0$$

is less than 1.
So, we proved

Theorem 7.5
$$\partial\rho/\partial t +_n \ div(\rho \times_n v) = \omega_3$$
We assume that all elements of this equality belong to W_n.

Corollary 7.6 *The probability of equality*

$$\partial\rho/\partial t +_n \ div(\rho \times_n v) = 0$$

is less than 1.

So, we get equation of continuity in Mathematics with Observers.
By **D18** we get

$$\partial\rho/\partial t +_n \rho \times_n div\mathbf{v} +_n (\mathbf{v}, \mathbf{grad}\rho) +_n \omega_4 = \omega_3$$

where ω_4 is a random variable depends on n and m.
So, the probability of equation

$$\partial\rho/\partial t +_n \rho \times_n div\mathbf{v} +_n (\mathbf{v}, \mathbf{grad}\rho) = 0$$

is less than 1.
So, we proved

Theorem 7.7

$$\partial\rho/\partial t +_n \rho \times_n div\boldsymbol{v} +_n (\boldsymbol{v}, \boldsymbol{grad}\rho) +_n \omega_4 = \omega_3$$

We assume that all elements of this equality belong to W_n.

Corollary 7.8 *The probability of equality*

$$\partial\rho/\partial t +_n \rho \times_n div\boldsymbol{v} +_n (\boldsymbol{v}, \boldsymbol{grad}\rho) = 0$$

is less than 1.

8

Observability and Euler Equation of Motion of the Fluid

We have pressure as a function $p(x, y, z)$, where

$$x \in [a, b], y \in [c, d], z \in [e, g]$$

$$\partial p/\partial x, \partial p/\partial y, \partial p/\partial z$$

exist and $\in W_n$ for any point (x, y, z) inside cube in the ideal fluid $[a, b] \times [c, d] \times [e, g]$, and

$$x, y, z, p, a, b, c, d, e, g \in W_n$$

The total force acting on this volume is the vector

$$-\left(\int_e^g \int_c^d (p(b, y, z) \times_n \Delta y) \times_n \Delta z -_n \int_e^g \int_c^d (p(a, y, z) \times_n \Delta y) \times_n \Delta z \right),$$

$$\int_e^g \int_a^b (p(x, d, z) \times_n \Delta x) \times_n \Delta z -_n \int_e^g \int_a^b (p(x, c, z) \times_n \Delta x) \times_n \Delta z,$$

$$\int_a^b \int_c^d (p(x, y, g) \times_n \Delta y) \times_n \Delta x -_n \int_a^b \int_c^d (p(x, y, e) \times_n \Delta y) \times_n \Delta x)$$

By **I15** we get

$$-\left(\int_e^g \int_c^d (p(b, y, z) \times_n \Delta y) \times_n \Delta z -_n \int_e^g \int_c^d (p(a, y, z) \times_n \Delta y) \times_n \Delta z \right),$$

$$\int_e^g \int_a^b (p(x, d, z) \times_n \Delta x) \times_n \Delta z -_n \int_e^g \int_a^b (p(x, c, z) \times_n \Delta x) \times_n \Delta z,$$

$$\int_a^b \int_c^d (p(x, y, g) \times_n \Delta y) \times_n \Delta x -_n \int_a^b \int_c^d (p(x, y, e) \times_n \Delta y) \times_n \Delta x) =$$

DOI: 10.1201/9781003175902-8

$$= -\int_e^g \int_c^d \int_a^b ((\mathbf{grad}p(x,y,z) \times_n \Delta x) \times_n \Delta y) \times_n \Delta z +_n \omega_5$$

where ω_5 is a random vector depends on n and m.

So, the probability of equation

$$-\left(\int_e^g \int_c^d (p(b,y,z) \times_n \Delta y) \times_n \Delta z -_n \int_e^g \int_c^d (p(a,y,z) \times_n \Delta y) \times_n \Delta z\right),$$

$$\int_e^g \int_a^b (p(x,d,z) \times_n \Delta x) \times_n \Delta z -_n \int_e^g \int_u^b (p(x,c,z) \times_n \Delta x) \times_n \Delta z,$$

$$\int_a^b \int_c^d (p(x,y,g) \times_n \Delta y) \times_n \Delta x -_n \int_a^b \int_c^d (p(x,y,e) \times_n \Delta y) \times_n \Delta x =$$

$$= -\int_e^g \int_c^d \int_a^b ((\mathbf{grad}p(x,y,z) \times_n \Delta x) \times_n \Delta y) \times_n \Delta z$$

is less than 1.

So, we proved

Theorem 8.1

$$-\left(\int_e^g \int_c^d (p(b,y,z) \times_n \Delta y) \times_n \Delta z -_n \int_e^g \int_c^d (p(a,y,z) \times_n \Delta y) \times_n \Delta z\right),$$

$$\int_e^g \int_a^b (p(x,d,z) \times_n \Delta x) \times_n \Delta z -_n \int_e^g \int_a^b (p(x,c,z) \times_n \Delta x) \times_n \Delta z,$$

$$\int_a^b \int_c^d (p(x,y,g) \times_n \Delta y) \times_n \Delta x -_n \int_a^b \int_c^d (p(x,y,e) \times_n \Delta y) \times_n \Delta x =$$

$$= -\int_e^g \int_c^d \int_a^b ((\boldsymbol{grad}p(x,y,z) \times_n \Delta x) \times_n \Delta y) \times_n \Delta z +_n \omega_5$$

We assume that all elements of this equality belong to W_n.

Corollary 8.2 *The probability of equality*

$$-\left(\int_e^g \int_c^d (p(b,y,z) \times_n \Delta y) \times_n \Delta z -_n \int_e^g \int_c^d (p(a,y,z) \times_n \Delta y) \times_n \Delta z\right),$$

$$\int_e^g \int_a^b (p(x,d,z) \times_n \Delta x) \times_n \Delta z -_n \int_e^g \int_a^b (p(x,c,z) \times_n \Delta x) \times_n \Delta z,$$

$$\int_a^b \int_c^d (p(x,y,g) \times_n \Delta y) \times_n \Delta x -_n \int_a^b \int_c^d (p(x,y,e) \times_n \Delta y) \times_n \Delta x) =$$

$$= -\int_e^g \int_c^d \int_a^b ((\boldsymbol{grad}p(x,y,z) \times_n \Delta x) \times_n \Delta y) \times_n \Delta z$$

is less than 1.

So, we can say the force

$$-\mathbf{grad}p +_n \omega_6$$

acts on unit volume of the fluid, where ω_6 is a random vector depends on n and m.

The equation of motion of the volume element in the fluid by Newton's second law is

$$\rho \times_n \frac{d\mathbf{v}}{dt} = -\mathbf{grad}p +_n \omega_6$$

where ρ is the mass per unit volume and $\frac{d\mathbf{v}}{dt}$ is the acceleration.
So, the probability of equation

$$\rho \times_n \frac{d\mathbf{v}}{dt} = -\mathbf{grad}p$$

is less than 1.
So, we proved

Theorem 8.3
$$\rho \times_n \frac{d\boldsymbol{v}}{dt} = -\boldsymbol{grad}p +_n \omega_6$$

We assume that all elements of this equality belong to W_n.

Corollary 8.4 *The probability of equality*

$$\rho \times_n \frac{d\boldsymbol{v}}{dt} = -\boldsymbol{grad}p$$

is less than 1.

By **D2** we get

$$\frac{d\mathbf{v}}{dt} = \partial\mathbf{v}/\partial t +_n (\mathbf{v},\nabla) \times_n \mathbf{v} +_n \omega_7$$

where ω_7 is a random vector depends on n and m.
So, the probability of equation

$$\frac{d\mathbf{v}}{dt} = \partial\mathbf{v}/\partial t +_n (\mathbf{v},\nabla) \times_n \mathbf{v}$$

is less than 1.
So, we proved

Theorem 8.5

$$\frac{d\boldsymbol{v}}{dt} = \partial \boldsymbol{v}/\partial t +_n (\boldsymbol{v}, \nabla) \times_n \boldsymbol{v} +_n \omega_7$$

We assume that all elements of this equality belong to W_n.

Corollary 8.6 *The probability of equality*

$$\frac{d\boldsymbol{v}}{dt} = \partial \boldsymbol{v}/\partial t +_n (\boldsymbol{v}, \nabla) \times_n \boldsymbol{v}$$

is less than 1.

So, we get

$$\rho \times_n (\partial \mathbf{v}/\partial t +_n (\mathbf{v}, \nabla) \times_n \mathbf{v} +_n \omega_7) = -\mathbf{grad}p +_n \omega_6$$

This is Euler equation of motion of the fluid in Mathematics with Observers.

So, probability of equation

$$\rho \times_n (\partial \mathbf{v}/\partial t +_n (\mathbf{v}, \nabla) \times_n \mathbf{v}) = -\mathbf{grad}p$$

is less than 1.

So, we proved

Theorem 8.7

$$\rho \times_n (\partial \boldsymbol{v}/\partial t +_n (\boldsymbol{v}, \nabla) \times_n \boldsymbol{v} +_n \omega_7) = -\boldsymbol{grad}p +_n \omega_6$$

We assume that all elements of this equality belong to W_n.

Corollary 8.8 *The probability of equality*

$$\rho \times_n (\partial \boldsymbol{v}/\partial t +_n (\boldsymbol{v}, \nabla) \times_n \boldsymbol{v}) = -\boldsymbol{grad}p$$

is less than 1.

If the fluid is in gravitational field, an additional force

$$\rho \times_n \mathbf{g}$$

where $\mathbf{g}$ is acceleration due to gravity, acts on any unit volume.
This force must be added to right side of equation

$$\rho \times_n \frac{d\mathbf{v}}{dt} = -\mathbf{grad}p +_n \omega_6$$

and Euler equation in Mathematics with Observers may be written as

$$\rho \times_n (\partial \mathbf{v}/\partial t +_n (\mathbf{v}, \nabla) \times_n \mathbf{v} +_n \omega_7) = -\mathbf{grad}p +_n \rho \times_n \mathbf{g} +_n \omega_6$$

And probability of equation

$$\rho \times_n (\partial \mathbf{v}/\partial t +_n (\mathbf{v}, \nabla) \times_n \mathbf{v}) = -\mathbf{grad}p +_n \rho \times_n \mathbf{g}$$

is less than 1.
So, we proved

Theorem 8.9

$$\rho \times_n (\partial v/\partial t +_n (v, \nabla) \times_n v +_n \omega_7) = -\boldsymbol{grad}p +_n \rho \times_n \boldsymbol{g} +_n \omega_6$$

We assume that all elements of this equality belong to W_n.

Corollary 8.10 *The probability of equality*

$$\rho \times_n (\partial v/\partial t +_n (v, \nabla) \times_n v) = -\boldsymbol{grad}p +_n \rho \times_n \boldsymbol{g}$$

is less than 1.

The deriving of the equations of motion we have taken no account of processes of energy dissipation which may occur in a moving fluid in consequence of internal friction (viscosity) in the fluid and heat exchange between different parts of it. So, we are considering the motions of fluids in which thermal conductivity and viscosity are unimportant, i.e. so called ideal fluids.

The absence of heat exchange between different parts of the fluid and the bodies adjoining it means that the motion is adiabatic throughout the fluid. In adiabatic motion the entropy of any particle of fluid remains constant as that particle moves about in space.

So, we have

$$ds/dt = 0$$

where s is the entropy per unit mass.
Let's consider equation (**TD2'**) for unit mass of fluid:

$$dh = T \times_n ds +_n V \times_n dp +_n \eta_{17}$$

where η_{17} is a random variable depends on n and m,
h is the heat function of unit mass, $V = \frac{1}{\rho}$ is the specific volume
(the probability of existing of $\frac{1}{\rho}$ is less than 1).
We assume that all elements of this equality belong to W_n.
So, the probability of equality

$$dh = T \times_n ds +_n V \times_n dp$$

is less than 1.
So, we proved

Theorem 8.11

$$dh = T \times_n ds +_n V \times_n dp +_n \eta_{17}$$

We assume that all elements of this equality belong to W_n.

Corollary 8.12 *The probability of equality*

$$dh = T \times_n ds +_n V \times_n dp$$

is less than 1.

Because $s = const$ we get

$$\rho \times_n dh = dp +_n \eta_{18}$$

where η_{18} is a random variable depends on n and m.
So, the probability of equality

$$\rho \times_n dh = dp$$

is less than 1.
So, we proved

Theorem 8.13

$$\rho \times_n dh = dp +_n \eta_{18}$$

We assume that all elements of this equality belong to W_n.

Corollary 8.14 *The probability of equality*

$$\rho \times_n dh = dp$$

is less than 1.

And we can write

$$\rho \times_n \mathbf{grad}h = \mathbf{grad}p +_n \eta_{19}$$

where η_{19} is a random vector depends on n and m.
So, the probability of equality

$$\rho \times_n \mathbf{grad}h = \mathbf{grad}p$$

is less than 1.
So, we proved

Theorem 8.15

$$\rho \times_n \boldsymbol{grad}h = \boldsymbol{grad}p +_n \eta_{19}$$

We assume that all elements of this equality belong to W_n.

Corollary 8.16 *The probability of equality*

$$\rho \times_n \boldsymbol{gradh} = \boldsymbol{gradp}$$

is less than 1.

We can rewrite now Euler equation of motion of the fluid in Mathematics with Observers:

$$\partial\mathbf{v}/\partial t +_n (\mathbf{v}, \nabla) \times_n \mathbf{v} +_n \omega_7 = -\mathbf{grad}h +_n \eta_{20}$$

where η_{20} is a random vector depends on n and m.
So, the probability of equality

$$\partial\mathbf{v}/\partial t +_n (\mathbf{v}, \nabla) \times_n \mathbf{v} = -\mathbf{grad}h$$

is less than 1.
So, we proved

Theorem 8.17

$$\partial\boldsymbol{v}/\partial t +_n (\boldsymbol{v}, \nabla) \times_n \boldsymbol{v} +_n \omega_7 = -\boldsymbol{grad}h +_n \eta_{20}$$

We assume that all elements of this equality belong to W_n.

Corollary 8.18 *The probability of equality*

$$\partial\boldsymbol{v}/\partial t +_n (\boldsymbol{v}, \nabla) \times_n \boldsymbol{v} = -\boldsymbol{grad}h$$

is less than 1.

In this case it is useful to form additional view of the Euler equation of motion of the fluid in Mathematics with Observers, in which it involves only the velocity.
By **(D7)** we get

$$\mathbf{grad}(\mathbf{v}, \mathbf{v}) = 2 \times_n (\mathbf{v}, \nabla)\mathbf{v} +_n 2 \times_n \mathbf{v} \times \mathbf{rot}\mathbf{v} +_n \xi_{74}$$

where ξ_{74} is a random vector depends on n and m.
So, the probability of equality

$$\mathbf{grad}(\mathbf{v}, \mathbf{v}) = 2 \times_n (\mathbf{v}, \nabla)\mathbf{v} +_n 2 \times_n \mathbf{v} \times \mathbf{rot}\mathbf{v}$$

is less than 1.
So, we proved

Theorem 8.19

$$\boldsymbol{grad}(\boldsymbol{v}, \boldsymbol{v}) = 2 \times_n (\boldsymbol{v}, \nabla)\boldsymbol{v} +_n 2 \times_n \boldsymbol{v} \times \boldsymbol{rot}\boldsymbol{v} +_n \xi_{74}$$

We assume that all elements of this equality belong to W_n.

Corollary 8.20 *The probability of equality*

$$grad(v, v) = 2 \times_n (v, \nabla)v +_n 2 \times_n v \times rotv$$

is less than 1.

And we can rewrite the Euler equation of motion of the fluid in Mathematics with Observers in this case as:

$$\partial \mathbf{v}/\partial t -_n \mathbf{v} \times \mathbf{rotv} = -\mathbf{grad}(h +_n 0.5(\mathbf{v}, \mathbf{v}) +_n \xi_{75}$$

where ξ_{75} is a random vector depends on n and m.
So, the probability of equality

$$\partial \mathbf{v}/\partial t -_n \mathbf{v} \times \mathbf{rotv} = -\mathbf{grad}(h +_n 0.5(\mathbf{v}, \mathbf{v})$$

is less than 1.
So, we proved

Theorem 8.21

$$\partial v/\partial t -_n v \times rotv = -grad(h +_n 0.5(v, v) +_n \xi_{75}$$

We assume that all elements of this equality belong to W_n.

Corollary 8.22 *The probability of equality*

$$\partial v/\partial t -_n v \times rotv = -grad(h +_n 0.5(v, v)$$

is less than 1.

If we take **rot** of both sides of the equation, we get

$$\partial(\mathbf{rotv})/\partial t = \mathbf{rot}(\mathbf{v} \times \mathbf{rotv}) +_n \xi_{76}$$

where ξ_{76} is a random vector depends on n and m.
So, the probability of equality

$$\partial(\mathbf{rotv})/\partial t = \mathbf{rot}(\mathbf{v} \times \mathbf{rotv})$$

is less than 1.
So, we proved

Theorem 8.23

$$\partial(rotv)/\partial t = rot(v \times rotv) +_n \xi_{76}$$

We assume that all elements of this equality belong to W_n.

Corollary 8.24 *The probability of equality*

$$\partial(\boldsymbol{rotv})/\partial t = \boldsymbol{rot}(\boldsymbol{v} \times \boldsymbol{rotv})$$

is less than 1.

Also we can rewrite now Euler equation of motion of the fluid in Mathematics with Observers in gravitational field:

$$\partial \mathbf{v}/\partial t +_n (\mathbf{v}, \nabla) \times_n \mathbf{v} +_n \omega_7 = -\mathbf{grad}h +_n \mathbf{g} +_n \eta_{21}$$

where η_{21} is the random vector depend on n and m.
So, the probability of equality

$$\partial \mathbf{v}/\partial t +_n (\mathbf{v}, \nabla) \times_n \mathbf{v} = -\mathbf{grad}h +_n \mathbf{g}$$

is less than 1.
So, we proved

Theorem 8.25

$$\partial \boldsymbol{v}/\partial t +_n (\boldsymbol{v}, \nabla) \times_n \boldsymbol{v} +_n \omega_7 = -\boldsymbol{grad}h +_n \boldsymbol{g} +_n \eta_{21}$$

We assume that all elements of this equality belong to W_n.

Corollary 8.26 *The probability of equality*

$$\partial \boldsymbol{v}/\partial t +_n (\boldsymbol{v}, \nabla) \times_n \boldsymbol{v} = -\boldsymbol{grad}h +_n \boldsymbol{g}$$

is less than 1.

Let's consider now a fluid at rest in uniform gravitational field. In this case the Euler equation in Mathematics with Observers takes the form:

$$\mathbf{grad}p = \rho \times_n \mathbf{g} +_n \omega_{22}$$

where η_{22} is a random vector depends on n and m.
So, the probability of equality

$$\mathbf{grad}p = \rho \times_n \mathbf{g}$$

is less than 1.
So, we proved

Theorem 8.27

$$\boldsymbol{grad}p = \rho \times_n \boldsymbol{g} +_n \omega_{22}$$

We assume that all elements of this equality belong to W_n.

Corollary 8.28 *The probability of equality*

$$\mathbf{grad}p = \rho \times_n \mathbf{g}$$

is less than 1.

Let's assume that the fluid is not only in mechanical equilibrium but also in thermal equilibrium. Then the temperature is the same in every point. Let's consider now the equation (**TD4'**) for unit mass of fluid:

$$d\Phi = V \times_n dp -_n s \times_n dT +_n \eta_{23}$$

where η_{23} is a random variable depends on n and m,
Φ is Gibbs free energy per unit mass.
We assume that all elements of this equality belong to W_n.
So, the probability of equality

$$d\Phi = V \times_n dp -_n s \times_n dT$$

is less than 1.
So, we proved

Theorem 8.29
$$d\Phi = V \times_n dp -_n s \times_n dT +_n \eta_{23}$$

We assume that all elements of this equality belong to W_n.

Corollary 8.30 *The probability of equality*

$$d\Phi = V \times_n dp -_n s \times_n dT$$

is less than 1.

For constant temperature

$$d\Phi = V \times_n dp +_n \eta_{23}$$

and

$$\rho \times_n d\Phi = dp +_n \eta_{24}$$

where η_{24} is a random variable depends on n and m.
So, the probabilities of equalities

$$d\Phi = V \times_n dp$$

and

$$\rho \times_n d\Phi = dp$$

are less than 1.
So, we proved

Theorem 8.31

$$d\Phi = V \times_n dp +_n \eta_{23}$$

and

$$\rho \times_n d\Phi = dp +_n \eta_{24}$$

We assume that all elements of these equalities belong to W_n.

Corollary 8.32 *The probabilities of equalities*

$$d\Phi = V \times_n dp$$

and

$$\rho \times_n d\Phi = dp$$

are less than 1.

And we can write the Euler equation in Mathematics with Observers for the fluid at rest in uniform gravitational field as follow:

$$\mathbf{grad}\Phi = \mathbf{g} +_n \omega_{25}$$

where η_{25} is a random vector depends on n and m.
So, the probability of equality

$$\mathbf{grad}\Phi = \mathbf{g}$$

is less than 1.
So, we proved

Theorem 8.33

$$\boldsymbol{grad}\Phi = \boldsymbol{g} +_n \omega_{25}$$

We assume that all elements of this equality belong to W_n.

Corollary 8.34 *The probability of equality*

$$\boldsymbol{grad}\Phi = \boldsymbol{g}$$

is less than 1.

9

Observability and Energy Flux and Moment Flux Equations

Let's take some volume element fixed in space. The energy of unit volume of ideal fluid is

$$\frac{1}{2} \times_n (\rho \times_n (\mathbf{v}, \mathbf{v})) +_n \rho \times_n \epsilon$$

where

$$\frac{1}{2} \times_n (\rho \times_n (\mathbf{v}, \mathbf{v}))$$

is a kinetic energy and

$$\rho \times_n \epsilon$$

is the internal energy, and ϵ is internal energy per unit mass. The change in this energy is

$$\partial/\partial t (\frac{1}{2} \times_n (\rho \times_n (\mathbf{v}, \mathbf{v})) +_n \rho \times_n \epsilon)$$

By **D1** we get

$$\partial/\partial t (\frac{1}{2} \times_n (\rho \times_n (\mathbf{v}, \mathbf{v})) +_n \rho \times_n \epsilon) =$$

$$= \partial/\partial t (\frac{1}{2} \times_n (\rho \times_n (\mathbf{v}, \mathbf{v}))) +_n \partial/\partial t (\rho \times_n \epsilon) +_n \omega_8$$

where ω_8 is a random variable depends on n and m.
So, the probability of equality

$$\partial/\partial t (\frac{1}{2} \times_n (\rho \times_n (\mathbf{v}, \mathbf{v})) +_n \rho \times_n \epsilon) =$$

$$= \partial/\partial t (\frac{1}{2} \times_n (\rho \times_n (\mathbf{v}, \mathbf{v}))) +_n \partial/\partial t (\rho \times_n \epsilon)$$

is less than 1.
So, we proved

DOI: 10.1201/9781003175902-9

Theorem 9.1

$$\partial/\partial t(\frac{1}{2} \times_n (\rho \times_n (\boldsymbol{v}, \boldsymbol{v})) +_n \rho \times_n \epsilon) =$$

$$= \partial/\partial t(\frac{1}{2} \times_n (\rho \times_n (\boldsymbol{v}, \boldsymbol{v}))) +_n \partial/\partial t(\rho \times_n \epsilon) +_n \omega_8$$

We assume that all elements of this equality belong to W_n.

Corollary 9.2 *The probability of equality*

$$\partial/\partial t(\frac{1}{2} \times_n (\rho \times_n (\boldsymbol{v}, \boldsymbol{v})) +_n \rho \times_n \epsilon) =$$

$$= \partial/\partial t(\frac{1}{2} \times_n (\rho \times_n (\boldsymbol{v}, \boldsymbol{v}))) +_n \partial/\partial t(\rho \times_n \epsilon)$$

is less than 1.

By **D17c** and **D3** let's calculate first

$$\partial/\partial t(\frac{1}{2} \times_n (\rho \times_n (\mathbf{v}, \mathbf{v}))) =$$

$$= (\frac{1}{2} \times_n (\mathbf{v}, \mathbf{v})) \times_n \partial\rho/\partial t +_n \omega_9 +_n (\frac{1}{2} \times_n \rho) \times_n (2 \times_n (\mathbf{v}, \partial v/\partial t)) +_n \omega_{10} =$$

$$= (\frac{1}{2} \times_n (\mathbf{v}, \mathbf{v})) \times_n \partial\rho/\partial t +_n \omega_9 +_n \rho \times_n (\mathbf{v}, \partial \mathbf{v}/\partial t) +_n \omega_{10} +_n \omega_{11}$$

where $\omega_9, \omega_{10}, \omega_{11}$ are the random variables depend on n and m.
So, the probability of equality

$$\partial/\partial t(\frac{1}{2} \times_n (\rho \times_n (\mathbf{v}, \mathbf{v}))) =$$

$$= (\frac{1}{2} \times_n (\mathbf{v}, \mathbf{v})) \times_n \partial\rho/\partial t +_n (\frac{1}{2} \times_n \rho) \times_n (2 \times_n (\mathbf{v}, \partial \mathbf{v}/\partial t)) =$$

$$= (\frac{1}{2} \times_n (\mathbf{v}, \mathbf{v})) \times_n \partial\rho/\partial t +_n \rho \times_n (\mathbf{v}, \partial \mathbf{v}/\partial t)$$

is less than 1.
So, we proved

Theorem 9.3

$$\partial/\partial t(\frac{1}{2} \times_n (\rho \times_n (\boldsymbol{v}, \boldsymbol{v}))) =$$

$$= (\frac{1}{2} \times_n (\boldsymbol{v}, \boldsymbol{v})) \times_n \partial\rho/\partial t +_n \omega_9 +_n (\frac{1}{2} \times_n \rho) \times_n (2 \times_n (\boldsymbol{v}, \partial \boldsymbol{v}/\partial t)) +_n \omega_{10} =$$

$$= (\frac{1}{2} \times_n (\boldsymbol{v}, \boldsymbol{v})) \times_n \partial\rho/\partial t +_n \omega_9 +_n \rho \times_n (\boldsymbol{v}, \partial \boldsymbol{v}/\partial t) +_n \omega_{10} +_n \omega_{11}$$

We assume that all elements of this equality belong to W_n.

Corollary 9.4 *The probability of equality*

$$\partial/\partial t(\frac{1}{2} \times_n (\rho \times_n (\boldsymbol{v}, \boldsymbol{v}))) =$$

$$= (\frac{1}{2} \times_n (\boldsymbol{v}, \boldsymbol{v})) \times_n \partial\rho/\partial t +_n (\frac{1}{2} \times_n \rho) \times_n (2 \times_n (\boldsymbol{v}, \partial \boldsymbol{v}/\partial t)) =$$

$$= (\frac{1}{2} \times_n (\boldsymbol{v}, \boldsymbol{v})) \times_n \partial\rho/\partial t +_n \rho \times_n (\boldsymbol{v}, \partial \boldsymbol{v}/\partial t)$$

is less than 1.

We get

$$\partial/\partial t(\frac{1}{2} \times_n (\rho \times_n (\mathbf{v}, \mathbf{v}))) = -(\frac{1}{2} \times_n (\mathbf{v}, \mathbf{v})) \times_n div(\rho \times_n \mathbf{v}) +_n \omega_{12} -_n$$

$$-_n(\mathbf{v}, \mathbf{grad}\ p) -_n (\rho \times_n \mathbf{v}, (\mathbf{v}, \nabla) \times_n \mathbf{v}) +_n \omega_{13}$$

where ω_{12}, ω_{13} are the random variables depend on n and m.
So, the probability of equality

$$\partial/\partial t(\frac{1}{2} \times_n (\rho \times_n (\mathbf{v}, \mathbf{v}))) = -(\frac{1}{2} \times_n (\mathbf{v}, \mathbf{v})) \times_n div(\rho \times_n \mathbf{v}) -_n$$

$$-_n(\mathbf{v}, \mathbf{grad}\ p) -_n (\rho \times_n \mathbf{v}, (\mathbf{v}, \nabla) \times_n \mathbf{v})$$

is less than 1.
So, we proved

Theorem 9.5

$$\partial/\partial t(\frac{1}{2} \times_n (\rho \times_n (\boldsymbol{v}, \boldsymbol{v}))) = -(\frac{1}{2} \times_n (\boldsymbol{v}, \boldsymbol{v})) \times_n div(\rho \times_n \boldsymbol{v}) +_n \omega_{12} -_n$$

$$-_n(\boldsymbol{v}, \boldsymbol{grad}\ p) -_n (\rho \times_n \boldsymbol{v}, (\boldsymbol{v}, \nabla) \times_n \boldsymbol{v}) +_n \omega_{13}$$

We assume that all elements of this equality belong to W_n.

Corollary 9.6 *The probability of equality*

$$\partial/\partial t(\frac{1}{2} \times_n (\rho \times_n (v, v))) = -(\frac{1}{2} \times_n (v, v)) \times_n div(\rho \times_n v) -_n$$

$$-_n(v, grad\ p) -_n (\rho \times_n v, (v, \nabla) \times_n v)$$

is less than 1.

Now we calculate

$$\partial/\partial t(\rho \times_n \epsilon) = \partial\rho/\partial t \times_n \epsilon +_n \rho \times_n \partial\epsilon/\partial t +_n \omega_{14} =$$

$$= -div(\rho \times_n v) \times_n \epsilon +_n \omega_{15} +_n \rho \times_n \partial\epsilon/\partial t +_n \omega_{14}$$

where ω_{14}, ω_{15} are the random variables depend on n and m. So, the probability of equality

$$\partial/\partial t(\rho \times_n \epsilon) = \partial\rho/\partial t \times_n \epsilon +_n \rho \times_n \partial\epsilon/\partial t =$$

$$= -div(\rho \times_n v) \times_n \epsilon +_n \rho \times_n \partial\epsilon/\partial t$$

is less than 1.
So, we proved

Theorem 9.7

$$\partial/\partial t(\rho \times_n \epsilon) = \partial\rho/\partial t \times_n \epsilon +_n \rho \times_n \partial\epsilon/\partial t +_n \omega_{14} =$$

$$= -div(\rho \times_n v) \times_n \epsilon +_n \omega_{15} +_n \rho \times_n \partial\epsilon/\partial t +_n \omega_{14}$$

We assume that all elements of this equality belong to W_n.

Corollary 9.8 *The probability of equality*

$$\partial/\partial t(\rho \times_n \epsilon) = \partial\rho/\partial t \times_n \epsilon +_n \rho \times_n \partial\epsilon/\partial t =$$

$$= -div(\rho \times_n v) \times_n \epsilon +_n \rho \times_n \partial\epsilon/\partial t$$

is less than 1.

So, we get

$$\partial/\partial t(\frac{1}{2} \times_n (\rho \times_n (v, v)) +_n \rho \times_n \epsilon) =$$

$$= -(\frac{1}{2} \times_n (v, v)) \times_n div(\rho \times_n v) -_n (v, grad\ p) -_n (\rho \times_n v, (v, \nabla) \times_n v) -_n$$

$$-_n div(\rho \times_n \mathbf{v}) \times_n \epsilon +_n \rho \times_n \partial\epsilon/\partial t +_n \omega_{16}$$

where ω_{16} is a random variable depends on n and m.

This equation we can name as Mathematics with Observers energy flux equation.

So, the probability of equality

$$\partial/\partial t(\frac{1}{2} \times_n (\rho \times_n (\mathbf{v}, \mathbf{v})) +_n \rho \times_n \epsilon) =$$

$$= -(\frac{1}{2} \times_n (\mathbf{v}, \mathbf{v})) \times_n div(\rho \times_n \mathbf{v}) -_n (\mathbf{v}, \mathbf{grad}\ p) -_n (\rho \times_n \mathbf{v}, (\mathbf{v}, \nabla) \times_n \mathbf{v}) -_n$$

$$-_n div(\rho \times_n \mathbf{v}) \times_n \epsilon +_n \rho \times_n \partial\epsilon/\partial t$$

is less than 1.

So, we proved

Theorem 9.9

$$\partial/\partial t(\frac{1}{2} \times_n (\rho \times_n (v, v)) +_n \rho \times_n \epsilon) =$$

$$= -(\frac{1}{2} \times_n (v, v)) \times_n div(\rho \times_n v) -_n (v, \mathbf{grad}\ p) -_n (\rho \times_n v, (v, \nabla) \times_n v) -_n$$

$$-_n div(\rho \times_n v) \times_n \epsilon +_n \rho \times_n \partial\epsilon/\partial t +_n \omega_{16}$$

We assume that all elements of this equality belong to W_n.

Corollary 9.10 *The probability of equality*

$$\partial/\partial t(\frac{1}{2} \times_n (\rho \times_n (v, v)) +_n \rho \times_n \epsilon) =$$

$$= -(\frac{1}{2} \times_n (v, v)) \times_n div(\rho \times_n v) -_n (v, \mathbf{grad}\ p) -_n (\rho \times_n v, (v, \nabla) \times_n v) -_n$$

$$-_n div(\rho \times_n v) \times_n \epsilon +_n \rho \times_n \partial\epsilon/\partial t$$

is less than 1.

The momentum of unit volume of ideal fluid is

$$\rho \times_n \mathbf{v}$$

The rate of this momentum change is

$$\partial(\rho \times_n \mathbf{v})/\partial t$$

By **D2** we have

$$\partial(\rho \times_n v_i)/\partial t = \rho \times_n \partial v_i/\partial t +_n \partial \rho/\partial t \times_n v_i +_n \omega_{16}^i; i = 1, 2, 3$$

where $\omega_{16}^1, \omega_{16}^2, \omega_{16}^3$ are the random variables depend on n and m.
So, the probability of equality

$$\partial(\rho \times_n v_i)/\partial t = \rho \times_n \partial v_i/\partial t +_n \partial \rho/\partial t \times_n v_i; i = 1, 2, 3$$

is less than 1.
So, we proved

Theorem 9.11

$$\partial(\rho \times_n v_i)/\partial t = \rho \times_n \partial v_i/\partial t +_n \partial \rho/\partial t \times_n v_i +_n \omega_{16}^i; i = 1, 2, 3$$

We assume that all elements of this equality belong to W_n.

Corollary 9.12 *The probability of equality*

$$\partial(\rho \times_n v_i)/\partial t = \rho \times_n \partial v_i/\partial t +_n \partial \rho/\partial t \times_n v_i; i = 1, 2, 3$$

is less than 1.

We get

$$\partial \rho/\partial t = -\partial(\rho \times_n v_k)/\partial x_k +_n \omega_{17}; k = 1, 2, 3$$

where ω_{17} is the random variable depends on n and m.
So, the probability of equality

$$\partial \rho/\partial t = -\partial(\rho \times_n v_k)/\partial x_k; k = 1, 2, 3$$

is less than 1.
So, we proved

Theorem 9.13

$$\partial \rho/\partial t = -\partial(\rho \times_n v_k)/\partial x_k +_n \omega_{17}; k = 1, 2, 3$$

We assume that all elements of this equality belong to W_n.

Corollary 9.14 *The probability of equality*

$$\partial \rho/\partial t = -\partial(\rho \times_n v_k)/\partial x_k; k = 1, 2, 3$$

is less than 1.

Let's note we consider everything here in not-moving coordinate system

$$x = x_1, y = x_2, z = x_3$$

and notation

$$\partial(\rho \times_n v_k)/\partial x_k$$

means summation by k from 1 to 3.

We get for $i = 1, 2, 3$

$$\rho \times_n (\partial v_i/\partial t +_n v_k \times_n \partial v_i/\partial x_k) = -\partial p/\partial x_i +_n \omega_{18}^i$$

where $\omega_{18}^1, \omega_{18}^2, \omega_{18}^3$ are the random variables depend on n and m. So, the probability of equality

$$\rho \times_n (\partial v_i/\partial t +_n v_k \times_n \partial v_i/\partial x_k) = -\partial p/\partial x_i$$

is less than 1.

So, we proved

Theorem 9.15

$$\rho \times_n (\partial v_i/\partial t +_n v_k \times_n \partial v_i/\partial x_k) = -\partial p/\partial x_i +_n \omega_{18}^i$$

We assume that all elements of this equality belong to W_n.

Corollary 9.16 *The probability of equality*

$$\rho \times_n (\partial v_i/\partial t +_n v_k \times_n \partial v_i/\partial x_k) = -\partial p/\partial x_i$$

is less than 1.

And again let's note that notation

$$v_k \times_n \partial v_i/\partial x_k$$

means summation by k from 1 to 3.

So, we have

$$\partial(\rho \times_n v_i)/\partial t = -\rho \times_n (v_k \times_n \partial v_i/\partial x_k) -_n \partial p/\partial x_i -_n$$

$$-_n v_i \times_n \partial(\rho \times_n v_k)/\partial x_k +_n \omega_{19}^i =$$

$$= -\partial p/\partial x_i -_n \partial/\partial x_k((\rho \times_n v_i) \times_n v_k) +_n \omega_{19}^i +_n \omega_{20}^i$$

where $\omega_{19}^1, \omega_{19}^2, \omega_{19}^3, \omega_{20}^1, \omega_{20}^2, \omega_{20}^3$ are the random variables depend on n and m.

So, the probability of equality

$$\partial(\rho \times_n v_i)/\partial t = -\rho \times_n (v_k \times_n \partial v_i/\partial x_k) -_n \partial p/\partial x_i -_n$$

$$-_n v_i \times_n \partial(\rho \times_n v_k)/\partial x_k =$$

$$= -\partial p/\partial x_i -_n \partial/\partial x_k((\rho \times_n v_i) \times_n v_k)$$

is less than 1.

So, we proved

Theorem 9.17

$$\partial(\rho \times_n v_i)/\partial t = -\rho \times_n (v_k \times_n \partial v_l/\partial x_k) -_n \partial p/\partial x_i -_n$$

$$-_n v_i \times_n \partial(\rho \times_n v_k)/\partial x_k +_n \omega_{19}^i =$$

$$= -\partial p/\partial x_i -_n \partial/\partial x_k((\rho \times_n v_i) \times_n v_k) +_n \omega_{19}^i +_n \omega_{20}^i$$

We assume that all elements of this equality belong to W_n.

Corollary 9.18 *The probability of equality*

$$\partial(\rho \times_n v_i)/\partial t = -\rho \times_n (v_k \times_n \partial v_i/\partial x_k) -_n \partial p/\partial x_i -_n$$

$$-_n v_i \times_n \partial(\rho \times_n v_k)/\partial x_k =$$

$$= -\partial p/\partial x_i -_n \partial/\partial x_k((\rho \times_n v_i) \times_n v_k)$$

is less than 1.

And again note expressions

$$\rho \times_n (v_k \times_n \partial v_i/\partial x_k), v_i \times_n \partial(\rho \times_n v_k)/\partial x_k, \partial/\partial x_k((\rho \times_n v_i) \times_n v_k)$$

mean summation by k from 1 to 3 for any $i = 1, 2, 3$.

We can rewrite

$$\partial p/\partial x_i = \delta_{ik} \times_n \partial p/\partial x_k$$

where we get summation by k and

$$(\delta_{ik}) = \begin{bmatrix} 1 & 0 & 0 \\ 0 & 1 & 0 \\ 0 & 0 & 1 \end{bmatrix}$$

and after that we can write down

$$\partial(\rho \times_n v_i)/\partial t = -\partial \Pi_{ik}/\partial x_k +_n \omega_{21}^i$$

where as usual expression

$$\partial \Pi_{ik}/\partial x_k$$

means summation by k from 1 to 3 for any $i = 1, 2, 3$, and where

$$\Pi_{ik} = p \times_n \delta_{ik} +_n (\rho \times_n v_i) \times_n v_k$$

$i, k = 1, 2, 3$
and $\omega_{21}^1, \omega_{21}^2, \omega_{21}^3$ are the random variables depend on n and m.
So, the probability of equality

$$\partial(\rho \times_n v_i)/\partial t = -\partial \Pi_{ik}/\partial x_k$$

is less than 1.
So, we proved

Theorem 9.19

$$\partial(\rho \times_n v_i)/\partial t = -\partial \Pi_{ik}/\partial x_k +_n \omega_{21}^i$$

We assume that all elements of this equality belong to W_n.

Corollary 9.20 *The probability of equality*

$$\partial(\rho \times_n v_i)/\partial t = -\partial \Pi_{ik}/\partial x_k$$

is less than 1.

To see the meaning of Π_{ik} let's consider

$$\partial/\partial t\left(\int_e^g \int_c^d \int_a^b ((\rho \times_n v_i \times_n \Delta x_1) \times_n \Delta x_2) \times_n \Delta x_3\right) =$$

$$= -\int_e^g \int_c^d \int_a^b ((\partial \Pi_{ik}/\partial x_k \times_n \Delta x_1) \times_n \Delta x_2) \times_n \Delta x_3 +_n \omega_{22}^i$$

where we consider some volume

$$[a, b] \times [c, d] \times [e, g]$$

and $\omega_{22}^1, \omega_{22}^2, \omega_{22}^3$ are the random variables depend on n and m.
So, the probability of equality

$$\partial/\partial t\left(\int_e^g \int_c^d \int_a^b ((\rho \times_n v_i \times_n \Delta x_1) \times_n \Delta x_2) \times_n \Delta x_3\right) =$$

$$= -\int_e^g \int_c^d \int_a^b ((\partial \Pi_{ik}/\partial x_k \times_n \Delta x_1) \times_n \Delta x_2) \times_n \Delta x_3$$

is less than 1.
So, we proved

Theorem 9.21

$$\partial/\partial t \left(\int_e^g \int_c^d \int_a^b ((\rho \times_n v_i \times_n \Delta x_1) \times_n \Delta x_2) \times_n \Delta x_3 \right) =$$

$$= -\int_e^g \int_c^d \int_a^b ((\partial \Pi_{ik}/\partial x_k \times_n \Delta x_1) \times_n \Delta x_2) \times_n \Delta x_3 +_n \omega_{22}^i$$

We assume that all elements of this equality belong to W_n.

Corollary 9.22 *The probability of equality*

$$\partial/\partial t \left(\int_e^g \int_c^d \int_a^b ((\rho \times_n v_i \times_n \Delta x_1) \times_n \Delta x_2) \times_n \Delta x_3 \right) =$$

$$= -\int_e^g \int_c^d \int_a^b ((\partial \Pi_{ik}/\partial x_k \times_n \Delta x_1) \times_n \Delta x_2) \times_n \Delta x_3$$

is less than 1.

And expression

$$\int_e^g \int_c^d \int_a^b ((\partial \Pi_{ik}/\partial x_k \times_n \Delta x_1) \times_n \Delta x_2 \times_n \Delta x_3$$

means summation by k from 1 to 3 for any $i = 1, 2, 3$.
By **I13** we get

$$\partial/\partial t \left(\int_e^g \int_c^d \int_a^b ((\rho \times_n v_i \times_n \Delta x_1) \times_n \Delta x_2) \times_n \Delta x_3 \right) =$$

$$= -\left(\int_e^g \int_c^d (\Pi_{i1}(b, x_2, x_3) \times_n \Delta x_2) \times_n \Delta x_3 -_n \right.$$

$$-_n \int_e^g \int_c^d (\Pi_{i1}(a, x_2, x_3) \times_n \Delta x_2) \times_n \Delta x_3 +_n$$

$$+_n \int_e^g \int_a^b (\Pi_{i2}(x_1, d, x_3) \times_n \Delta x_1) \times_n \Delta x_3 -_n$$

$$-_n \int_e^g \int_a^b (\Pi_{i2}(x_1, c, x_3) \times_n \Delta x_1) \times_n \Delta x_3 +_n$$

$$+_n \int_a^b \int_c^d (\Pi_{i3}(x_1, x_2, g) \times_n \Delta x_2) \times_n \Delta x_1 -_n$$

$$-_n \int_a^b \int_c^d (\Pi_{i3}(x_1, y_2, e) \times_n \Delta x_2) \times_n \Delta x_1) +_n \omega_{23}^i$$

where $\omega_{23}^1, \omega_{23}^2, \omega_{23}^3$ are the random variables depend on n and m, $i = 1, 2, 3$.

The left hand side of this equation is the rate of change of the ith component of the momentum contained in considered volume. The right hand side is the amount of the momentum flowing out through the bounding surface in unit time. So, Π_{ik} is the momentum flux density matrix with accuracy up to several random variables.

And the probability of equality

$$\partial/\partial t\left(\int_e^g \int_c^d \int_a^b ((\rho \times_n v_i \times_n \Delta x_1) \times_n \Delta x_2) \times_n \Delta x_3\right) =$$

$$= -\left(\int_e^g \int_c^d (\Pi_{i1}(b, x_2, x_3) \times_n \Delta x_2) \times_n \Delta x_3 -_n\right.$$

$$-_n \int_e^g \int_c^d (\Pi_{i1}(a, x_2, x_3) \times_n \Delta x_2) \times_n \Delta x_3 +_n$$

$$+_n \int_e^g \int_a^b (\Pi_{i2}(x_1, d, x_3) \times_n \Delta x_1) \times_n \Delta x_3 -_n$$

$$-_n \int_e^g \int_a^b (\Pi_{i2}(x_1, c, x_3) \times_n \Delta x_1) \times_n \Delta x_3 +_n$$

$$+_n \int_a^b \int_c^d (\Pi_{i3}(x_1, x_2, g) \times_n \Delta x_2) \times_n \Delta x_1 -_n$$

$$-_n \int_a^b \int_c^d (\Pi_{i3}(x_1, y_2, e) \times_n \Delta x_2) \times_n \Delta x_1\right)$$

is less than 1.

So, we proved

Theorem 9.23

$$\partial/\partial t\left(\int_e^g \int_c^d \int_a^b ((\rho \times_n v_i \times_n \Delta x_1) \times_n \Delta x_2) \times_n \Delta x_3\right) =$$

$$= -\left(\int_e^g \int_c^d (\Pi_{i1}(b, x_2, x_3) \times_n \Delta x_2) \times_n \Delta x_3 -_n\right.$$

$$-_n \int_e^g \int_c^d (\Pi_{i1}(a, x_2, x_3) \times_n \Delta x_2) \times_n \Delta x_3 +_n$$

$$+_n \int_e^g \int_a^b (\Pi_{i2}(x_1, d, x_3) \times_n \Delta x_1) \times_n \Delta x_3 -_n$$

$$-_n \int_e^g \int_a^b (\Pi_{i2}(x_1, c, x_3) \times_n \Delta x_1) \times_n \Delta x_3 +_n$$

$$+_n \int_a^b \int_c^d (\Pi_{i3}(x_1, x_2, g) \times_n \Delta x_2) \times_n \Delta x_1 -_n$$

$$-_n \int_a^b \int_c^d (\Pi_{i3}(x_1, y_2, e) \times_n \Delta x_2) \times_n \Delta x_1) +_n \omega_{23}^i$$

We assume that all elements of this equality belong to W_n.

Corollary 9.24 *The probability of equality*

$$\partial/\partial t \left(\int_e^g \int_c^d \int_a^b ((\rho \times_n v_i \times_n \Delta x_1) \times_n \Delta x_2) \times_n \Delta x_3 \right) =$$

$$= -\left(\int_e^g \int_c^d (\Pi_{i1}(b, x_2, x_3) \times_n \Delta x_2) \times_n \Delta x_3 -_n \right.$$

$$-_n \int_e^g \int_c^d (\Pi_{i1}(a, x_2, x_3) \times_n \Delta x_2) \times_n \Delta x_3 +_n$$

$$+_n \int_e^g \int_a^b (\Pi_{i2}(x_1, d, x_3) \times_n \Delta x_1) \times_n \Delta x_3 -_n$$

$$-_n \int_e^g \int_a^b (\Pi_{i2}(x_1, c, x_3) \times_n \Delta x_1) \times_n \Delta x_3 +_n$$

$$+_n \int_a^b \int_c^d (\Pi_{i3}(x_1, x_2, g) \times_n \Delta x_2) \times_n \Delta x_1 -_n$$

$$-_n \int_a^b \int_c^d (\Pi_{i3}(x_1, y_2, e) \times_n \Delta x_2) \times_n \Delta x_1)$$

is less than 1.

So, Π_{ik} is the momentum flux density matrix with accuracy up to several random variables. In classic Fluid Mechanics Π_{ik} is called the momentum flux density tensor (see [8]). Let's note that in Mathematics with Observers Π_{ik} is not a tensor in classic linear algebra sense, but only matrix in the chosen coordinate system. Let's consider now this situation in details.

We have the following statements for classic linear algebra:

T1. In m-dimensional linear space any m linear independent vectors form the basis of this space.

T2. Each vector in m-dimensional linear space is a linear combination of m linear independent vectors (i.e. basis vectors).

T3. For any new basis with vectors $\mathbf{e'}_i$ and any old basis with vectors $\mathbf{e}_i$ transition matrix always exists;

T4. In Euclidean space for each vector there is parallel unit vector.

Let's consider now Mathematics with Observers point of view.

Let's start from **T1** and **T2**.

And we consider vectors $\mathbf{a} = (3, 0, 0)$, $\mathbf{b} = (0, 3, 0)$, $\mathbf{c} = (0, 0, 3)$.

So, $m = 3$, $\mathbf{a}, \mathbf{b}, \mathbf{c} \in E_3 W_n$ and these three vectors are linear independent vectors.

And as we know there are no $\alpha, \beta.\gamma \in W_n$ such that

$$\alpha \times_n \mathbf{a} +_n \beta \times_n \mathbf{b} +_n \gamma \times_n \mathbf{c} = \mathbf{i}$$

or

$$\alpha \times_n \mathbf{a} +_n \beta \times_n \mathbf{b} +_n \gamma \times_n \mathbf{c} = \mathbf{j}$$

or

$$\alpha \times_n \mathbf{a} +_n \beta \times_n \mathbf{b} +_n \gamma \times_n \mathbf{c} = \mathbf{k}$$

Sure, if we take
$$\mathbf{e}_1 = \mathbf{i}, \mathbf{e}_2 = \mathbf{j}, \mathbf{e}_3 = \mathbf{k}$$
then each vector in $E_3 W_n$ is a linear combination of basis vectors $\mathbf{i}, \mathbf{j}, \mathbf{k}$.

So, statements **T1** and **T2** are correct in $E_3 W_n$ with some probability less than 1. We have the same situation in $E_m W_n$ for any $m \geq 3$.

So, we proved

Theorem 9.25 *For any $E_m W_n$ with $m \geq 3$ the probability of any m linear independent vectors form the basis of this space, and each vector of this space space is their linear combination is less than 1.*

Let's consider now **T3**.

If we take "old" system of 3 linear independent vectors $\mathbf{a}, \mathbf{b}, \mathbf{c} \in E_3 W_n$

$$\mathbf{a} = (3, 0, 0), \mathbf{b} = (0, 3, 0), \mathbf{c} = (0, 0, 3)$$

and "new" system

$$\mathbf{i}, \mathbf{j}, \mathbf{k}$$

transition matrix from "old system" to "new system" does not exist. Let's note "inverse" transition matrix from "new system" to "old system" exists.

And if we take old basis in $E_3 W_n$

$$\mathbf{e}_1 = (2, 0, 0), \mathbf{e}_2 = (0, 2, 0), \mathbf{e}_3 = (0, 0, 2)$$

and new basis

$$\mathbf{i}, \mathbf{j}, \mathbf{k}$$

transition matrix A exists:

$$A = \begin{bmatrix} 0.5 & 0 & 0 \\ 0 & 0.5 & 0 \\ 0 & 0 & 0.5 \end{bmatrix}$$

So, statement **T3** is correct in $E_3 W_n$ with some probability less than 1. We have the same situation in $E_m W_n$ for any $m \geq 3$.

So, we proved

Theorem 9.26 *If we take "old" system of m linear independent vectors and "new" system of m linear independent vectors in $E_m W_n$ with $m \geq 3$ then the probability of the transition matrix existence is less than 1.*

Finally, let's consider **T4**.

If we take vector $\mathbf{a} = (a_1, a_2, a_3) \in E_3 W_n$ with $a_1 = a_2 = a_3 = 1$, then

$$\alpha \times_n \mathbf{a} = (\alpha, \alpha, \alpha)$$

For $n = 2$ and

$$\alpha \in \pm 0.50, \pm 0.51, \pm 0.52, \pm 0.53, \pm 0.54, \pm 0.55, \pm 0.56, \pm 0.57, \pm 0.58, \pm 0.59$$

we have

$$\alpha^2 +_n \alpha^2 +_n \alpha^2 = 0.75 < 1$$

For

$$\alpha \in \pm 0.60, \pm 0.61, \pm 0.62, \pm 0.63, \pm 0.64, \pm 0.65, \pm 0.66, \pm 0.67, \pm 0.68, \pm 0.69$$

we have

$$\alpha^2 +_n \alpha^2 +_n \alpha^2 = 1.08 > 1$$

That means that such α with $|\alpha \times_n \mathbf{a}| = 1$ doesn't exist. From another side, vectors $\mathbf{i}, \mathbf{j}, \mathbf{k}$ have a length 1. So, statement **T4** is correct in $E_3 W_n$ with some probability less than 1.

So, we proved

Theorem 9.27 *The probability of in $E_m W_n$ each vector is parallel some unit vector is less than 1.*

As we know tensors in classic linear algebra are geometric objects with linear relations between them (see [8]). The components of a tensor change when we change the basis of the vector space. Each tensor has a transformation law that details how the components of the tensor respond to a change of basis. The transformation law for an order $p + q$ tensor with p contravariant indices and q covariant indices is

$$\hat{T}^{i'_1, \ldots, i'_p}_{j'_1, \ldots, j'_q} = \left(R^{-1}\right)^{i'_1}_{i_1} \ldots \left(R^{-1}\right)^{i'_p}_{i_p} R^{j_1}_{j'_1} \ldots R^{j_q}_{j'_q} T^{i_1, \ldots, i_p}_{j_1, \ldots, j_q}$$

Here the primed indices denote tensor components in the new coordinates, and the unprimed indices denote the tensor components in the old coordinates. Such a tensor called as a (p, q)-tensor.

And we get

Theorem 9.28 *Tensors of type (p, q) in classic linear algebra are not tensors in Mathematics with Observers. They are only tensors with some probability less than 1.*

10

Observability and Incompressible Fluids

Let's assume that the density of flow of liquids or gases is invariable, i.e constant throughout the volume of the fluid and throughout its motion – it is named the incompressible flow.

Euler equation in Mathematics with Observers does not change:

$$\rho \times_n (\partial \mathbf{v}/\partial t +_n (\mathbf{v}, \nabla) \times_n \mathbf{v} +_n \omega_7) = -\mathbf{grad}p +_n \rho \times_n \mathbf{g} +_n \omega_6$$

(only in this case $\rho = const$).

In this case because $\rho = const$ we can take equation, in which it involves only the velocity:

$$\partial(\mathbf{rotv})/\partial t = \mathbf{rot}(\mathbf{v} \times \mathbf{rotv}) +_n \xi_{76}$$

where ξ_{76} is a random vector depends on n and m.

So, the probability of equation

$$\partial(\mathbf{rotv})/\partial t = \mathbf{rot}(\mathbf{v} \times \mathbf{rotv})$$

is less than 1.

So, we proved

Theorem 10.1

$$\partial(\boldsymbol{rotv})/\partial t = \boldsymbol{rot}(\boldsymbol{v} \times \boldsymbol{rotv}) +_n \xi_{76}$$

We assume that all elements of this equality belong to W_n.

Corollary 10.2 *The probability of equality*

$$\partial(\boldsymbol{rotv})/\partial t = \boldsymbol{rot}(\boldsymbol{v} \times \boldsymbol{rotv})$$

is less than 1.

But equation of continuity in Mathematics with Observers becomes more simple:

$$\rho \times_n div\mathbf{v} +_n \omega_4 = \omega_3$$

The equations are particularly simple for potential flow (i.e $\mathbf{rotv} = \xi_{77}$)

DOI: 10.1201/9781003175902-10

of an incompressible fluid (ξ_{77} is a random vector depends on n and m). So, the probability of equation

$$\mathbf{rotv} = \mathbf{0}$$

is less than 1.
So, we proved

Theorem 10.3

$$\boldsymbol{rotv} = \xi_{77}$$

We assume that all elements of this equality belong to W_n.

Corollary 10.4 *The probability of equality*

$$\boldsymbol{rotv} = \boldsymbol{0}$$

is less than 1.

That means the Euler equation in Mathematics with Observers is satisfied with corresponding ξ_{77}. And if we take

$$\mathbf{v} = \mathbf{grad}\phi$$

we get from equation of continuity in Mathematics with Observers the following:

$$\Delta\phi = \xi_{78}$$

where ξ_{78} is a random variable depends on n and m, Δ is Laplace operator of potential ϕ.
So, the probability of equation

$$\Delta\phi = 0$$

is less than 1.
So, we proved

Theorem 10.5

$$\Delta\phi = \xi_{78}$$

We assume that all elements of this equality belong to W_n.

Corollary 10.6 *The probability of equality*

$$\Delta\phi = 0$$

is less than 1.

This equation must be supplemented by boundary conditions at the surfaces where the fluid meets solid bodies. Let's consider the situation of plane flow, i.e. the velocity distribution in a moving fluid depends on only two coordinates (for example x, y), and the velocity is everywhere parallel to the xy-plane.

We get in this case

$$div\mathbf{v} = \partial v_x/\partial x +_n \partial v_y/\partial y = \xi_{79}$$

where ξ_{79} is a random variable depends on n and m.

So, the probability of equation

$$div\mathbf{v} = \partial v_x/\partial x +_n \partial v_y/\partial y = 0$$

is less than 1.

So, we proved

Theorem 10.7

$$div\boldsymbol{v} = \partial v_x/\partial x +_n \partial v_y/\partial y = \xi_{79}$$

We assume that all elements of this equality belong to W_n.

Corollary 10.8 *The probability of equality*

$$div\boldsymbol{v} = \partial v_x/\partial x +_n \partial v_y/\partial y = 0$$

is less than 1.

We get

$$v_x = \partial\psi/\partial y +_n \xi_{80}$$

$$v_y = -\partial\psi/\partial x +_n \xi_{81}$$

where ξ_{80}, ξ_{81} are the random variables depend on n and m.

So, the probabilities of equations

$$v_x = \partial\psi/\partial y$$

$$v_y = -\partial\psi/\partial x$$

are less than 1.

So, we proved

Theorem 10.9

$$v_x = \partial\psi/\partial y +_n \xi_{80}$$
$$v_y = -\partial\psi/\partial x +_n \xi_{81}$$

We assume that all elements of these equalities belong to W_n.

Corollary 10.10 *The probabilities of equalities*

$$v_x = \partial\psi/\partial y$$

$$v_y = -\partial\psi/\partial x$$

are less than 1.

The function $\psi = \psi(x, y)$ we will call the stream function.
This function has to satisfy the following equation:

$$\partial(\Delta\psi)/\partial t -_n \partial\psi/\partial x \times_n \partial(\Delta\psi)/\partial y +_n \partial\psi/\partial y \times_n \partial(\Delta\psi)/\partial x = \xi_{82}$$

where ξ_{82} is a random variable depends on n and m.
So, the probability of equation

$$\partial(\Delta\psi)/\partial t -_n \partial\psi/\partial x \times_n \partial(\Delta\psi)/\partial y +_n \partial\psi/\partial y \times_n \partial(\Delta\psi)/\partial x = 0$$

is less than 1.
So, we proved

Theorem 10.11

$$\partial(\Delta\psi)/\partial t -_n \partial\psi/\partial x \times_n \partial(\Delta\psi)/\partial y +_n \partial\psi/\partial y \times_n \partial(\Delta\psi)/\partial x = \xi_{82}$$

We assume that all elements of this equality belong to W_n.

Corollary 10.12 *The probability of equality*

$$\partial(\Delta\psi)/\partial t -_n \partial\psi/\partial x \times_n \partial(\Delta\psi)/\partial y +_n \partial\psi/\partial y \times_n \partial(\Delta\psi)/\partial x = 0$$

is less than 1.

Now we have to come back and consider the set CW_n of complex numbers in Mathematics with Observers. Let's consider special case – the set **U (1)** of all complex numbers z with $|z| = 1$. In classic linear algebra this is a Lie group of all rotations of the unit circle on real plane. What do we have now in CW_n?
Let's take $n = 2$ and $z_1 = 0.61 +_2 i0.82$; $z_2 = 0.69 +_2 i0.86$.

$$(0.61 +_2 i0.82) \times_2 (0.61 -_2 i0.82) = 0.36 +_2 0.64 = 1$$

i.e.

$$z_1^{-1} = 0.61 -_2 i0.82; |z_1^{-1}| = |z_1| = 1$$

$$(0.69 +_2 i0.86) \times_2 (0.69 -_2 i0.86) = 0.36 +_2 0.64 = 1$$

i.e.

$$z_2^{-1} = 0.69 -_2 i0.86; |z_2^{-1}| = |z_2| = 1$$

So, $z_1, z_2, z_1^{-1}, z_2^{-1} \in \mathbf{U}$ (1).
If we take again $n = 2$ and $z_1 = 0.61 +_2 i0.82; z_2 = 0.69 +_2 i0.86$,
we get

$$(0.61 +_2 i0.82) \times_2 (0.63 -_2 i0.87) = 0.36 +_2 0.64 = 1$$

i.e.

$$(z_1^{-1})_1 = 0.63 -_2 i0.87; |(z_1^{-1})_1| = |z_1| = 1$$

$$(0.69 +_2 i0.86) \times_2 (0.65 -_2 i0.80) = 0.36 +_2 0.64 = 1$$

i.e.

$$(z_2^{-1})_1 = 0.65 -_2 i0.80; |(z_2^{-1})_1| = |z_2| = 1$$

So, $z_1 = 0.61 +_2 i0.82; z_2 = 0.69 +_2 i0.86$ have more than one
inverse numbers (by multiplication). Generally speaking z_1, z_2
have 100 different inverse numbers each from W_m- observer
point of view, $m \geq 3$.
We have

$$z_3 = z_1 \times_2 z_2 = -0.28 +_2 i0.96$$

$$(-0.28 +_2 i0.96) \times_2 (-0.28 -_2 i0.96) = 0.04 +_2 0.81 = 0.85 \neq 1$$

$$|z_3| = \sqrt{0.04 +_2 0.81} = \sqrt{0.85}$$

and $\sqrt{0.85}$ does not exist, because

$$0.99^2 = 0.81; 1.00^2 = 1$$

So,

$$z_3 \notin \mathbf{U}\ (1)$$

and the set $\mathbf{U}$ (1) is not a group in Mathematics with Observers. Group
definition's conditions take a place here with some probability less than 1.
Returning now to incompressible fluid we have

$$v_x = \partial\phi/\partial x = \partial\psi/\partial y +_n \xi_{80}$$

$$v_y = \partial\phi/\partial y = -\partial\psi/\partial x +_n \xi_{81}$$

So, up to random variables these relations mean Cauchy-Riemann conditions (see [2]) for a complex function

$$\omega = \phi +_n i\psi$$

to be an analytical function of the complex argument

$$z = x +_n iy$$

We have

$$\frac{d\omega}{dz} = \partial\phi/\partial x +_n i\partial\psi/\partial x +_n \xi_{83} = v_x -_n v_y +_n \xi_{84}$$

where $\xi_{83}, \xi_{84} \in CW_n$ are the complex random variables depend on n and m.

So, the probability of equation

$$\frac{d\omega}{dz} = \partial\phi/\partial x +_n i\partial\psi/\partial x = v_x -_n v_y$$

is less than 1.

So, we proved

Theorem 10.13

$$\frac{d\omega}{dz} = \partial\phi/\partial x +_n i\partial\psi/\partial x +_n \xi_{83} = v_x -_n v_y +_n \xi_{84}$$

We assume that all elements of this equality belong to W_n.

Corollary 10.14 *The probability of equality*

$$\frac{d\omega}{dz} = \partial\phi/\partial x +_n i\partial\psi/\partial x = v_x -_n v_y$$

is less than 1.

We'll call the function ω the complex potential in Mathematics with Observers, and $\frac{d\omega}{dz}$ – the complex velocity.

We get

$$\frac{d\omega}{dz} = |\mathbf{v}| \times_n u +_n \xi_{85}$$

where ξ_{85} is a complex random variable depends on n and m,
$u \in \mathbf{U}\,(1)$.

So, the probability of equation

$$\frac{d\omega}{dz} = |\mathbf{v}| \times_n u$$

is less than 1.

So, we proved

Theorem 10.15

$$\frac{d\omega}{dz} = |v| \times_n u +_n \xi_{85}$$

We assume that all elements of this equality belong to W_n.

Corollary 10.16 *The probability of equality*

$$\frac{d\omega}{dz} = |v| \times_n u$$

is less than 1.

And note:

1. The probability of $|\mathbf{v}|$ existing is less than 1.
2. The probability of u existing is less than 1.

11

Observability and Navier-Stokes Equations

In this section we consider Navier-Stokes equations and classic problem of solutions of this system (see [1]) from point of view Mathematics with Observers. Now we are going from ideal fluids to viscous fluids.

We have Mathematics with Observers Euler equation for $i = 1, 2, 3$

$$\partial(\rho \times_n v_i)/\partial t = -\partial \Pi_{ik}/\partial x_k +_n \omega_{21}^i$$

where Π_{ik} is the momentum flux density matrix with accuracy up to several random variables. This momentum flux represents a completely reversible transfer of momentum due simply to the mechanical transport of the different particles of fluid from place to place and to the pressure forces acting in the fluid. The viscosity (internal friction) causes another, irreversible, transfer of momentum from points where the velocity is large to those where it is small. The equation of motion of a viscous fluid may therefore be obtained by adding to "ideal" momentum flux a term $-\sigma_{ik}^{visc}$ which gives the irreversible "viscous" transfer of momentum in the fluid.

Thus we write the momentum flux density matrix in a viscous fluid in the form

$$\Pi_{ik} = p \times_n \delta_{ik} +_n (\rho \times_n v_i) \times_n v_k -_n \sigma_{ik}^{visc} = -\sigma_{ik} +_n (\rho \times_n v_i) \times_n v_k$$

where

$$\sigma_{ik} = -p \times_n \delta_{ik} +_n \sigma_{ik}^{visc}$$

We will call σ_{ik} a stress matrix, and σ_{ik}^{visc} a viscous stress matrix.

Following by classical physics with experimental basis we can find general expression for σ_{ik}^{visc}:

$$\sigma_{ik}^{visc} = \alpha \times_n (\partial v_i/\partial x_k +_n \partial v_k/\partial x_i +_n \beta \times_n (\delta_{ik} \times_n \partial v_l/\partial x_l) +_n \gamma \times_n (\delta_{ik} \times_n \partial v_l/\partial x_l)$$

where α, β, γ are independent on a velocity and the functions of pressure and temperature,

$$\alpha > 0, \gamma > 0$$

and where

$$\partial v_l/\partial x_l = \partial v_1/\partial x_1 +_n \partial v_2/\partial x_2 +_n \partial v_3/\partial x_3 = div\mathbf{v}$$

The value of parameter β is defined from condition:

DOI: 10.1201/9781003175902-11

$$\partial v_i/\partial x_k +_n \partial v_k/\partial x_i +_n \beta \times_n (\delta_{ik} \times_n \partial v_l/\partial x_l)$$

has the property of vanishing (or being nearest to zero) on taking the sum of components with $i = k$. The equation of motion of a viscous fluid can now be obtained by simply adding the expression

$$\partial \sigma_{ik}^{visc}/\partial x_k$$

to the right hand side of Euler equation in Mathematics with Observers.

$$\rho \times_n (\partial v_i/\partial t +_n v_k \times_n \partial v_i/\partial x_k) = -\partial p/\partial x_i +_n \omega_{24}^i$$

where we have summation by k, $i = 1, 2, 3$ and $\omega_{24}^1, \omega_{24}^2, \omega_{24}^3$ are the random variables depend on n and m.
So, the probability of equation

$$\rho \times_n (\partial v_i/\partial t +_n v_k \times_n \partial v_i/\partial x_k) = -\partial p/\partial x_i$$

is less than 1.
So, we proved

Theorem 11.1

$$\rho \times_n (\partial v_i/\partial t +_n v_k \times_n \partial v_i/\partial x_k) = -\partial p/\partial x_i +_n \omega_{24}^i$$

We assume that all elements of this equality belong to W_n.

Corollary 11.2 *The probability of equality*

$$\rho \times_n (\partial v_i/\partial t +_n v_k \times_n \partial v_i/\partial x_k) = -\partial p/\partial x_i$$

is less than 1.

We get

$$\rho \times_n (\partial v_i/\partial t +_n v_k \times_n \partial v_i/\partial x_k) = -\partial p/\partial x_i +_n \omega_{24}^i +_n$$

$$+_n \partial/\partial x_k(\alpha \times_n (\partial v_i/\partial x_k +_n \partial v_k/\partial x_i +_n \beta \times_n \delta_{ik} \times_n \partial v_l/\partial x_l)) +_n$$

$$+_n \partial/\partial x_i(\gamma \times_n \partial v_l/\partial x_l) +_n \omega_{25}^i$$

where we have summation by k, $i = 1, 2, 3$ and $\omega_{25}^1, \omega_{25}^2, \omega_{25}^3$ are the random variables depend on n and m.
So, the probability of equation

$$\rho \times_n (\partial v_i/\partial t +_n v_k \times_n \partial v_i/\partial x_k) = -\partial p/\partial x_i +_n$$

$$+_n \partial/\partial x_k (\alpha \times_n (\partial v_i/\partial x_k +_n \partial v_k/\partial x_i +_n \beta \times_n \delta_{ik} \times_n \partial v_l/\partial x_l))$$
$$+_n \partial/\partial x_i (\gamma \times_n \partial v_l/\partial x_l)$$

is less than 1.

So, we proved

Theorem 11.3

$$\rho \times_n (\partial v_i/\partial t +_n v_k \times_n \partial v_i/\partial x_k) = -\partial p/\partial x_i +_n \omega_{24}^i +_n$$

$$+_n \partial/\partial x_k (\alpha \times_n (\partial v_i/\partial x_k +_n \partial v_k/\partial x_i +_n \beta \times_n \delta_{ik} \times_n \partial v_l/\partial x_l)) +_n$$

$$+_n \partial/\partial x_i (\gamma \times_n \partial v_l/\partial x_l) +_n \omega_{25}^i$$

We assume that all elements of this equality belong to W_n.

Corollary 11.4 *The probability of equality*

$$\rho \times_n (\partial v_i/\partial t +_n v_k \times_n \partial v_i/\partial x_k) = -\partial p/\partial x_i +_n$$

$$+_n \partial/\partial x_k (\alpha \times_n (\partial v_i/\partial x_k +_n \partial v_k/\partial x_i +_n \beta \times_n \delta_{ik} \times_n \partial v_l/\partial x_l))$$
$$+_n \partial/\partial x_i (\gamma \times_n \partial v_l/\partial x_l)$$

is less than 1.

If the parameters α, β, γ do not change noticeably in the fluid, they may be considered as the constants. In this case we have

$$\rho \times_n (\partial \mathbf{v}/\partial t +_n (\mathbf{v}, \nabla) \times_n \mathbf{v}) = -\mathbf{grad} p +_n \alpha \times_n \Delta \mathbf{v} +_n$$

$$+_n (\alpha +_n \alpha \times_n \beta +_n \gamma) \times_n \mathbf{grad}(div v) +_n \omega_{26}$$

where ω_{26} is a random vector depends on n and m, $\omega_{26} = (\omega_{26}^1, \omega_{26}^2, \omega_{26}^3)$, and $\omega_{26}^1, \omega_{26}^2, \omega_{26}^3$ are the random variables depend on n and m, and

$$\Delta = \partial^2/\partial x_1^2 +_n \partial^2/\partial x_2^2 +_n \partial^2/\partial x_3^2$$

is a Laplace transformation.

This is Navier-Stokes equation in Mathematics with Observers.

So, the probability of equation

$$\rho \times_n (\partial \mathbf{v}/\partial t +_n (\mathbf{v}, \nabla) \times_n \mathbf{v}) = -\mathbf{grad} p +_n \alpha \times_n \Delta \mathbf{v} +_n$$

$$+_n (\alpha +_n \alpha \times_n \beta +_n \gamma) \times_n \mathbf{grad}(div v)$$

is less than 1.

So, we proved

Theorem 11.5

$$\rho \times_n (\partial v/\partial t +_n (v, \nabla) \times_n v) = -\boldsymbol{grad}p +_n \alpha \times_n \Delta v +_n$$

$$+_n (\alpha +_n \alpha \times_n \beta +_n \gamma) \times_n \boldsymbol{grad}(divv) +_n \omega_{26}$$

We assume that all elements of this equality belong to W_n.

Corollary 11.6 *The probability of equality*

$$\rho \times_n (\partial v/\partial t +_n (v, \nabla) \times_n v) = -\boldsymbol{grad}p +_n \alpha \times_n \Delta v +_n$$

$$+_n (\alpha +_n \alpha \times_n \beta +_n \gamma) \times_n \boldsymbol{grad}(divv)$$

is less than 1.

If the fluid may be regarded as incompressible, i.e.

$$\rho = const$$

we get

$$\partial \rho/\partial t +_n \rho \times_n div\mathbf{v} +_n (\mathbf{v}, \mathbf{grad}\rho) +_n \omega_4 = \omega_3$$

and

$$\partial \rho/\partial t = 0, \mathbf{grad}\rho = 0$$

In this case

$$\rho \times_n div\mathbf{v} = \omega_{27}$$

where ω_{27} is a random variable depends on n and m, and

$$div\mathbf{v} = \omega_{28}$$

where ω_{28} is a random variable depends on n and m.
So, the probabilities of equations

$$\rho \times_n div\mathbf{v} = 0$$

$$div\mathbf{v} = 0$$

are less than 1.
So, we proved

Theorem 11.7

$$\rho \times_n divv = \omega_{27}$$

$$divv = \omega_{28}$$

We assume that all elements of these equalities belong to W_n.

Corollary 11.8 *The probabilities of equalities*

$$\rho \times_n div\boldsymbol{v} = 0$$

$$div\boldsymbol{v} = 0$$

are less than 1.

For incompressible fluid we get stress matrix

$$\sigma_{ik} = -p \times_n \delta_{ik} +_n \alpha \times_n (\partial v_i/\partial x_k +_n \partial v_k/\partial x_i)$$

and simplified Navier-Stokes equations in Mathematics with Observers:

$$\rho \times_n (\partial\mathbf{v}/\partial t +_n (\mathbf{v}, \nabla) \times_n \mathbf{v}) = -\mathbf{grad}p +_n \alpha \times_n \Delta\mathbf{v} +_n \omega_{29} +_n \omega_{26}$$

where ω_{29} is a random vector depends on n and m, $\omega_{29} = (\omega_{29}^1, \omega_{29}^2, \omega_{29}^3)$, and $\omega_{29}^1, \omega_{29}^2, \omega_{29}^3$ are the random variables depend on n and m. So, the probability of equation

$$\rho \times_n (\partial\mathbf{v}/\partial t +_n (\mathbf{v}, \nabla) \times_n \mathbf{v}) = -\mathbf{grad}p +_n \alpha \times_n \Delta\mathbf{v}$$

is less than 1.
So, we proved

Theorem 11.9

$$\rho \times_n (\partial\boldsymbol{v}/\partial t +_n (\boldsymbol{v}, \nabla) \times_n \boldsymbol{v}) = -\boldsymbol{grad}p +_n \alpha \times_n \Delta\boldsymbol{v} +_n \omega_{29} +_n \omega_{26}$$

We assume that all elements of this equality belong to W_n.

Corollary 11.10 *The probability of equality*

$$\rho \times_n (\partial\boldsymbol{v}/\partial t +_n (\boldsymbol{v}, \nabla) \times_n \boldsymbol{v}) = -\boldsymbol{grad}p +_n \alpha \times_n \Delta\boldsymbol{v}$$

is less than 1.

So, we get a system of equations:

$$\begin{cases} \rho \times_n (\partial\mathbf{v}/\partial t +_n (\mathbf{v}, \nabla) \times_n \mathbf{v}) = -\mathbf{grad}p +_n \alpha \times_n \Delta\mathbf{v} +_n \omega_{29} +_n \omega_{26} \\ div\mathbf{v} = \omega_{28} \end{cases}$$

So, the probability of the system of equations

$$\begin{cases} \rho \times_n (\partial\mathbf{v}/\partial t +_n (\mathbf{v}, \nabla) \times_n \mathbf{v}) = -\mathbf{grad}p +_n \alpha \times_n \Delta\mathbf{v} \\ div\mathbf{v} = 0 \end{cases}$$

is less than 1.
So, we proved

Theorem 11.11

$$\begin{cases} \rho \times_n (\partial v/\partial t +_n (v, \nabla) \times_n v) = -\mathbf{grad}p +_n \alpha \times_n \Delta v +_n \omega_{29} +_n \omega_{26} \\ div v = \omega_{28} \end{cases}$$

We assume that all elements of these equalities belong to W_n.

Corollary 11.12 *The probability of the system of equations*

$$\begin{cases} \rho \times_n (\partial v/\partial t +_n (v, \nabla) \times_n v) = -\mathbf{grad}p +_n \alpha \times_n \Delta v \\ div v = 0 \end{cases}$$

is less than 1.

If the fluid is in gravitational field first equation of this system may be rewritten as

$$\rho \times_n (\partial \mathbf{v}/\partial t +_n (\mathbf{v}, \nabla) \times_n \mathbf{v}) = -\mathbf{grad}p +_n \rho \times_n \mathbf{g} +_n \alpha \times_n \Delta \mathbf{v} +_n \omega_{29} +_n \omega_{26}$$

So, Navier-Stokes equations in Mathematics with Observers for incompressible fluid in gravitation field is represented as the following system:

$$\begin{cases} \rho \times_n (\partial \mathbf{v}/\partial t +_n (\mathbf{v}, \nabla) \times_n \mathbf{v}) = -\mathbf{grad}p +_n \rho \times_n \mathbf{g} +_n \alpha \times_n \Delta \mathbf{v} +_n \omega_{29} +_n \omega_{26} \\ div \mathbf{v} = \omega_{28} \end{cases}$$

So, the probability of the system of equations

$$\begin{cases} \rho \times_n (\partial \mathbf{v}/\partial t +_n (\mathbf{v}, \nabla) \times_n \mathbf{v}) = -\mathbf{grad}p +_n \rho \times_n \mathbf{g} +_n \alpha \times_n \Delta \mathbf{v} \\ div \mathbf{v} = 0 \end{cases}$$

is less than 1.
So, we proved

Theorem 11.13

$$\begin{cases} \rho \times_n (\partial v/\partial t +_n (v, \nabla) \times_n v) = -\mathbf{grad}p +_n \rho \times_n g +_n \alpha \times_n \Delta v +_n \omega_{29} +_n \omega_{26} \\ div v = \omega_{28} \end{cases}$$

We assume that all elements of these equalities belong to W_n.

Corollary 11.14 *The probability of the system of equations*

$$\begin{cases} \rho \times_n (\partial v/\partial t +_n (v, \nabla) \times_n v) = -\mathbf{grad}p +_n \rho \times_n g +_n \alpha \times_n \Delta v \\ div v = 0 \end{cases}$$

is less than 1.

We can rewrite Navier-Stokes equations in Mathematics with Observers for incompressible fluid in coordinates

$$\mathbf{x} = (x_1, x_2, x_3), t \in W_n, t \geq 0$$

as follows:

$$\begin{cases} \rho \times_n (\partial v_i/\partial t +_n \sum_{j=1}^3 {}^n v_j \times_n \partial v_i/\partial x_j) = -\partial p/\partial x_i +_n f_i(\mathbf{x}, t) +_n A_i \\ A_i = \alpha \times_n (\sum_{j=1}^3 {}^n \partial^2 v_i/\partial x_j^2) +_n \omega_{29}^i +_n \omega_{26}^i \\ div\mathbf{v} = \sum_{j=1}^3 {}^n \partial v_j/\partial x_j = \omega_{28} \\ i = 1, 2, 3 \\ \alpha > 0 \end{cases}$$

where we have unknown velocity vector

$$\mathbf{v}(\mathbf{x}, t) = (v_1(x_1, x_2, x_3, t), v_2(x_1, x_2, x_3, t), v_3(x_1, x_2, x_3, t)) \in E_3 W_n$$

and unknown function - pressure

$$p = p(\mathbf{x}, t) = p(x_1, x_2, x_3, t) \in W_n$$

We assume here that we have given vector

$$\mathbf{f}(\mathbf{x}, t) = (f_1(x_1, x_2, x_3, t), f_2(x_1, x_2, x_3, t), f_3(x_1, x_2, x_3, t)) \in E_3 W_n$$

which is externally applied force (e.g. gravity). And we assume that all elements of this system belong to W_n.

So, the probability of the system of equations

$$\begin{cases} \rho \times_n (\partial v_i/\partial t +_n \sum_{j=1}^3 {}^n v_j \times_n \partial v_i/\partial x_j) = -\partial p/\partial x_i +_n f_i(\mathbf{x}, t) +_n B_i \\ B_i = \alpha \times_n (\sum_{j=1}^3 {}^n \partial^2 v_i/\partial x_j^2) \\ div\mathbf{v} = \sum_{j=1}^3 {}^n \partial v_j/\partial x_j = 0 \\ i = 1, 2, 3 \\ \alpha > 0 \end{cases}$$

is less than 1.

So, we proved

Theorem 11.15

$$\begin{cases} \rho \times_n (\partial v_i/\partial t +_n \sum_{j=1}^3 {}^n v_j \times_n \partial v_i/\partial x_j) = -\partial p/\partial x_i +_n f_i(\mathbf{x}, t) +_n A_i \\ A_i = \alpha \times_n (\sum_{j=1}^3 {}^n \partial^2 v_i/\partial x_j^2) +_n \omega_{29}^i +_n \omega_{26}^i \\ div\boldsymbol{v} = \sum_{j=1}^3 {}^n \partial v_j/\partial x_j = \omega_{28} \\ i = 1, 2, 3 \\ \alpha > 0 \end{cases}$$

We assume that all elements of these equalities belong to W_n.

Corollary 11.16 *The probability of the system of equations*

$$\begin{cases} \rho \times_n (\partial v_i/\partial t +_n \sum_{j=1}^3 {}^n v_j \times_n \partial v_i/\partial x_j) = -\partial p/\partial x_i +_n f_i(\boldsymbol{x}, t) +_n B_i \\ B_i = \alpha \times_n (\sum_{j=1}^3 {}^n \partial^2 v_i/\partial x_j^2) \\ div\boldsymbol{v} = \sum_{j=1}^3 {}^n \partial v_j/\partial x_j = 0 \\ i = 1, 2, 3 \\ \alpha > 0 \end{cases}$$

is less than 1.

We consider here derivatives (and partial derivatives) in three different meanings – right derivatives, left derivatives and both-sides derivatives (or partial derivatives). So, this system of equations has three different meanings.

We have also initial conditions

$$\mathbf{v}(\mathbf{x}, 0) = \mathbf{v}^0(\mathbf{x})$$

where $\mathbf{v}^0(\mathbf{x})$ is a given divergence-free vector.

In right partial derivatives meaning we get a system

$$\begin{cases} \rho \times_n ((\partial v_i/\partial t)_+ +_n \sum_{j=1}^3 {}^n v_j \times_n (\partial v_i/\partial x_j)_+) = -(\partial p/\partial x_i)_+ +_n (F_i)_+ +_n C_i \\ C_i = \alpha \times_n (\sum_{j=1}^3 {}^n (\partial^2 v_i/\partial x_j^2)_+) \\ div\mathbf{v} = \sum_{j=1}^3 {}^n (\partial v_j/\partial x_j)_+ = \omega_{31} \\ i = 1, 2, 3 \\ \alpha > 0 \end{cases}$$

where

$$(F_i)_+ = f_i(\mathbf{x}, t) +_n \omega_{30}^i$$

and $\omega_{30}^i, \omega_{31}$ are the random variables depend on n and m.
Also

$$(\partial^2 v_i/\partial x_j^2)_+ = (\partial/\partial x_j)_+ (\partial v_i/\partial x_j)_+$$

In left partial derivatives meaning we get a system

$$\begin{cases} \rho \times_n ((\partial v_i/\partial t)_- +_n \sum_{j=1}^3 {}^n v_j \times_n (\partial v_i/\partial x_j)_-) = -(\partial p/\partial x_i)_- +_n (F_i)_- +_n D_i \\ D_i = \alpha \times_n (\sum_{j=1}^3 {}^n (\partial^2 v_i/\partial x_j^2)_-) \\ div\mathbf{v} = \sum_{j=1}^3 {}^n (\partial v_j/\partial x_j)_- = \omega_{32} \\ i = 1, 2, 3 \\ \alpha > 0 \end{cases}$$

where

$$(F_i)_- = f_i(\mathbf{x}, t) +_n \omega_{33}^i$$

and $\omega_{32}, \omega_{33}^i$ are the random variables depend on n and m.

Also

$$(\partial^2 v_i/\partial x_j^2)_- = (\partial/\partial x_j)_- (\partial v_i/\partial x_j)_-$$

In both-sides derivative meaning we consider original system

$$\begin{cases} \rho \times_n (\partial v_i/\partial t +_n \sum_{j=1}^3 {}^n v_j \times_n \partial v_i/\partial x_j) = -\partial p/\partial x_i +_n f_i(\mathbf{x}, t) +_n A_i \\ A_i = \alpha \times_n (\sum_{j=1}^3 {}^n \partial^2 v_i/\partial x_j^2) +_n \omega_{29}^i +_n \omega_{26}^i \\ div\mathbf{v} = \sum_{j=1}^3 {}^n \partial v_j/\partial x_j = \omega_{28} \\ i = 1, 2, 3 \\ \alpha > 0 \end{cases}$$

By **D7** we see

$$\frac{1}{2} \times_n \mathbf{grad}(\mathbf{v}, \mathbf{v}) = (\mathbf{v}, \nabla) \times_n \mathbf{v} +_n \mathbf{v} \times \mathbf{rot}\ \mathbf{v} +_n \omega_{34}$$

where ω_{34} is a random vector depends on n and m.
So, the probability of equation

$$\frac{1}{2} \times_n \mathbf{grad}(\mathbf{v}, \mathbf{v}) = (\mathbf{v}, \nabla) \times_n \mathbf{v} +_n \mathbf{v} \times \mathbf{rot}\ \mathbf{v}$$

is less than 1.
So, we proved

Theorem 11.17

$$\frac{1}{2} \times_n \boldsymbol{grad}(\boldsymbol{v}, \boldsymbol{v}) = (\boldsymbol{v}, \nabla) \times_n \boldsymbol{v} +_n \boldsymbol{v} \times \boldsymbol{rot}\ \boldsymbol{v} +_n \omega_{34}$$

We assume that all elements of this equality belong to W_n.

Corollary 11.18 *The probability of equality*

$$\frac{1}{2} \times_n \boldsymbol{grad}(\boldsymbol{v}, \boldsymbol{v}) = (\boldsymbol{v}, \nabla) \times_n \boldsymbol{v} +_n \boldsymbol{v} \times \boldsymbol{rot}\ \boldsymbol{v}$$

is less than 1.

So,

$$(\mathbf{v}, \nabla) \times_n \mathbf{v} = \frac{1}{2} \times_n \mathbf{grad}(\mathbf{v}, \mathbf{v}) -_n \mathbf{v} \times \mathbf{rot}\ \mathbf{v} -_n \omega_{34}$$

We get

$$\rho \times_n (\partial \mathbf{v}/\partial t +_n (\mathbf{v}, \nabla) \times_n \mathbf{v} +_n \omega_7) = -\mathbf{grad}p +_n \omega_6$$

We have

$$\rho \times_n (\partial \mathbf{v}/\partial t +_n \frac{1}{2} \times_n \mathbf{grad}(\mathbf{v}, \mathbf{v}) -_n \mathbf{v} \times \mathbf{rot} \ \mathbf{v} -_n \omega_{34} +_n \omega_7) = -\mathbf{grad}p +_n \omega_6$$

And

$$\rho \times_n (\partial \mathbf{v}/\partial t -_n \mathbf{v} \times \mathbf{rot} \ \mathbf{v}) = -\mathbf{grad}p -_n \rho \times_n (\frac{1}{2} \times_n \mathbf{grad}(\mathbf{v}, \mathbf{v})) +_n \omega_{35}$$

where ω_{35} is a random vector depends on n and m.
So, the probability of equation

$$\rho \times_n (\partial \mathbf{v}/\partial t -_n \mathbf{v} \times \mathbf{rot} \ \mathbf{v}) = -\mathbf{grad}p -_n \rho \times_n (\frac{1}{2} \times_n \mathbf{grad}(\mathbf{v}, \mathbf{v}))$$

is less than 1.
So, we proved

Theorem 11.19

$$\rho \times_n (\partial v/\partial t -_n v \times rot \ v) = -gradp -_n \rho \times_n (\frac{1}{2} \times_n grad(v, v)) +_n \omega_{35}$$

We assume that all elements of this equality belong to W_n.

Corollary 11.20 *The probability of equality*

$$\rho \times_n (\partial v/\partial t -_n v \times rot \ v) = -gradp -_n \rho \times_n (\frac{1}{2} \times_n grad(v, v))$$

is less than 1.

We are considering now incompressible fluid, i.e.

$$\rho = const$$

Using **D1, D2, D3, D4, D5, D6**, and **D8** and taking **rot** we get

$$\partial/\partial t(\mathbf{rot} \ \mathbf{v}) -_n \mathbf{rot} \ (\mathbf{v} \times \mathbf{rot} \ \mathbf{v}) = \omega_{36}$$

where ω_{36} is a random vector depends on n and m.
So, the probability of equation

$$\partial/\partial t(\mathbf{rot} \ \mathbf{v}) -_n \mathbf{rot} \ (\mathbf{v} \times \mathbf{rot} \ \mathbf{v}) = \mathbf{0}$$

is less than 1.
So, we proved

Theorem 11.21

$$\partial/\partial t(\textbf{rot } v) -_n \textbf{rot } (v \times \textbf{rot } v) = \omega_{36}$$

We assume that all elements of this equality belong to W_n.

Corollary 11.22 *The probability of equality*

$$\partial/\partial t(\textbf{rot } v) -_n \textbf{rot } (v \times \textbf{rot } v) = \textbf{0}$$

is less than 1.

We get

$$(\mathbf{v}, \nabla) \times_n \mathbf{v} = \frac{1}{2} \times_n \textbf{grad}(\mathbf{v}, \mathbf{v}) -_n \mathbf{v} \times \textbf{rot } \mathbf{v} -_n \omega_{34}$$

And we get the equation for incompressible fluid in gravitation field

$$\rho \times_n (\partial \mathbf{v}/\partial t +_n (\mathbf{v}, \nabla) \times_n \mathbf{v}) = -\textbf{grad}p +_n \rho \times_n \mathbf{g} +_n \alpha \times_n \Delta \mathbf{v} +_n \omega_{29} +_n \omega_{26}$$

We have

$$\rho \times_n (\partial \mathbf{v}/\partial t +_n \frac{1}{2} \times_n \textbf{grad}(\mathbf{v}, \mathbf{v}) -_n \mathbf{v} \times \textbf{rot } \mathbf{v} -_n \omega_{34}) =$$

$$= -\textbf{grad}p +_n \rho \times_n \mathbf{g} +_n \alpha \times_n \Delta \mathbf{v} +_n \omega_{29} +_n \omega_{26}$$

And

$$\rho \times_n (\partial \mathbf{v}/\partial t -_n \mathbf{v} \times \textbf{rot } \mathbf{v}) =$$

$$= -\textbf{grad}p +_n \rho \times_n \mathbf{g} +_n \alpha \times_n \Delta \mathbf{v} -_n \rho \times_n (\frac{1}{2} \times_n \textbf{grad}(\mathbf{v}, \mathbf{v})) +_n \omega_{37}$$

where ω_{37} is a random vector depends on n and m.
So, the probability of equation

$$\rho \times_n (\partial \mathbf{v}/\partial t -_n \mathbf{v} \times \textbf{rot } \mathbf{v}) =$$

$$= -\textbf{grad}p +_n \rho \times_n \mathbf{g} +_n \alpha \times_n \Delta \mathbf{v} -_n \rho \times_n (\frac{1}{2} \times_n \textbf{grad}(\mathbf{v}, \mathbf{v}))$$

is less than 1.
So, we proved

Theorem 11.23

$$\rho \times_n (\partial v/\partial t -_n v \times rot\ v) =$$

$$= -grad p +_n \rho \times_n g +_n \alpha \times_n \Delta v -_n \rho \times_n (\frac{1}{2} \times_n grad(v, v)) +_n \omega_{37}$$

We assume that all elements of this equality belong to W_n.

Corollary 11.24 *The probability of equality*

$$\rho \times_n (\partial v/\partial t -_n v \times rot\ v) =$$

$$= -grad p +_n \rho \times_n g +_n \alpha \times_n \Delta v -_n \rho \times_n (\frac{1}{2} \times_n grad(v, v))$$

is less than 1.

Again using **D1, D2, D3, D4, D5, D6, D8**, and relation

$$\rho = const$$

and taking **rot** we get

$$\rho \times_n (\partial/\partial t(rot\ v) -_n rot\ (v \times rot\ v)) =$$

$$= \rho \times_n rot\ g +_n \alpha \times_n rot\Delta v +_n \omega_{38}$$

where

$$\omega_{38} = (\omega_{38}^1, \omega_{38}^2, \omega_{38}^3)$$

is a random vector with random coordinates depend on n and m. So, the probability of equation

$$\rho \times_n (\partial/\partial t(rot\ v) -_n rot\ (v \times rot\ v)) =$$

$$= \rho \times_n rot\ g +_n \alpha \times_n rot\Delta v$$

is less than 1.
So, we proved

Theorem 11.25

$$\rho \times_n (\partial/\partial t(rot\ v) -_n rot\ (v \times rot\ v)) =$$

$$= \rho \times_n rot\ g +_n \alpha \times_n rot\Delta v +_n \omega_{38}$$

We assume that all elements of this equality belong to W_n.

Corollary 11.26 *The probability of equality*

$$\rho \times_n (\partial/\partial t(\boldsymbol{rot\ v}) -_n \boldsymbol{rot}\ (\boldsymbol{v} \times \boldsymbol{rot\ v})) =$$

$$= \rho \times_n \boldsymbol{rot\ g} +_n \alpha \times_n \boldsymbol{rot}\Delta\boldsymbol{v}$$

is less than 1.

So, Navier-Stokes equations in Mathematics with Observers for incompressible fluid in gravitation field is represented as the following system:

$$\begin{cases} \rho \times_n (\partial/\partial t(\mathbf{rot\ v}) -_n \mathbf{rot}\ (\mathbf{v} \times \mathbf{rot\ v})) = \rho \times_n \mathbf{rot\ g} +_n \alpha \times_n \mathbf{rot}\Delta\mathbf{v} +_n \omega_{38} \\ div\mathbf{v} = \omega_{28} \\ \alpha > 0 \end{cases}$$

with initial conditions

$$\mathbf{v}(\mathbf{x}, 0) = \mathbf{v}^0(\mathbf{x})$$

where $\mathbf{v}^0(\mathbf{x})$ is a given divergence-free vector.
So, the probability of the system of equations

$$\begin{cases} \rho \times_n (\partial/\partial t(\mathbf{rot\ v}) -_n \mathbf{rot}\ (\mathbf{v} \times \mathbf{rot\ v})) = \rho \times_n \mathbf{rot\ g} +_n \alpha \times_n \mathbf{rot}\Delta\mathbf{v} \\ div\mathbf{v} = 0 \\ \alpha > 0 \end{cases}$$

is less than 1.
So, we proved

Theorem 11.27

$$\begin{cases} \rho \times_n (\partial/\partial t(\boldsymbol{rot\ v}) -_n \boldsymbol{rot}\ (\boldsymbol{v} \times \boldsymbol{rot\ v})) = \rho \times_n \boldsymbol{rot\ g} +_n \alpha \times_n \boldsymbol{rot}\Delta\boldsymbol{v} +_n \omega_{38} \\ div\boldsymbol{v} = \omega_{28} \\ \alpha > 0 \end{cases}$$

We assume that all elements of these equalities belong to W_n.

Corollary 11.28 *The probability of the system of equations*

$$\begin{cases} \rho \times_n (\partial/\partial t(\boldsymbol{rot\ v}) -_n \boldsymbol{rot}\ (\boldsymbol{v} \times \boldsymbol{rot\ v})) = \rho \times_n \boldsymbol{rot\ g} +_n \alpha \times_n \boldsymbol{rot}\Delta\boldsymbol{v} \\ div\boldsymbol{v} = 0 \\ \alpha > 0 \end{cases}$$

is less than 1.

Now we rewrite this system of equations for incompressible fluid in coordinates

$$\mathbf{x} = (x_1, x_2, x_3), t \in W_n, t \geq 0$$

with unknown velocity vector

$$\mathbf{v}(\mathbf{x}, t) = (v_1(x_1, x_2, x_3, t), v_2(x_1, x_2, x_3, t), v_3(x_1, x_2, x_3, t)) \in E_3 W_n$$

and with given initial conditions

$$\mathbf{v}(\mathbf{x}, 0) = \mathbf{v}^0(\mathbf{x})$$

So,

$$\mathbf{rot}\ \mathbf{v} = (\partial v_3/\partial x_2 -_n \partial v_2/\partial x_3, \partial v_1/\partial x_3 -_n \partial v_3/\partial x_1, \partial v_2/\partial x_1 -_n \partial v_1/\partial x_2)$$

$$\partial/\partial t(\mathbf{rot}\ \mathbf{v}) =$$

$$= (\partial/\partial t(\partial v_3/\partial x_2) -_n \partial/\partial t(\partial v_2/\partial x_3), \partial/\partial t(\partial v_1/\partial x_3) -_n \partial/\partial t(\partial v_3/\partial x_1),$$

$$\partial/\partial t(\partial v_2/\partial x_1) -_n \partial/\partial t(\partial v_1/\partial x_2)) +_n (\omega_{39}^1, \omega_{39}^2, \omega_{39}^3) =$$

$$= (\partial^2 v_3/\partial x_2 \partial t -_n \partial^2 v_2/\partial x_3 \partial t, \partial^2 v_1/\partial x_3 \partial t -_n \partial^2 v_3/\partial x_1 \partial t,$$

$$\partial^2 v_2/\partial x_1 \partial t -_n \partial^2 v_1/\partial x_2 \partial t) +_n (\omega_{39}^1, \omega_{39}^2, \omega_{39}^3) +_n (\omega_{39}^4, \omega_{39}^5, \omega_{39}^6)$$

where

$$\omega_{39_1} = (\omega_{39}^1, \omega_{39}^2, \omega_{39}^3)$$

and

$$\omega_{39_2} = (\omega_{39}^4, \omega_{39}^5, \omega_{39}^6)$$

are the random vectors with random coordinates depend on n and m.

$$\mathbf{v} \times \mathbf{rot v} =$$

$$= (v_2 \times_n \partial v_2/\partial x_1 -_n v_2 \times_n \partial v_1/\partial x_2 +_n v_3 \times_n \partial v_3/\partial x_1 -_n v_3 \times_n \partial v_1/\partial x_3,$$

$$-v_1 \times_n \partial v_2/\partial x_1 +_n v_1 \times_n \partial v_1/\partial x_2 +_n v_3 \times_n \partial v_3/\partial x_2 -_n v_3 \times_n \partial v_2/\partial x_3,$$

$$-v_1 \times_n \partial v_3/\partial x_1 +_n v_1 \times_n \partial v_1/\partial x_3 -_n v_2 \times_n \partial v_3/\partial x_2 +_n v_2 \times_n \partial v_2/\partial x_3) +_n$$

$$+_n(\omega_{40}^1, \omega_{40}^2, \omega_{40}^3)$$

where $\omega_{40} = (\omega_{40}^1, \omega_{40}^2, \omega_{40}^3)$ is a random vector with random coordinates depend on n and m.

$$\mathbf{rot}\ (\mathbf{v} \times \mathbf{rot}\ \mathbf{v}) = (R_1, R_2, R_3) +_n (\omega_{41}^1, \omega_{41}^2, \omega_{41}^3)$$

where $\omega_{41} = (\omega_{41}^1, \omega_{41}^2, \omega_{41}^3)$ is a random vector with random coordinates depend on n and m.

Calculations of coordinates R_1, R_2, R_3 see below:

$$R_1 = \partial/\partial x_2(-v_1 \times_n \partial v_3/\partial x_1 +_n v_1 \times_n \partial v_1/\partial x_3 -_n v_2 \times_n \partial v_3/\partial x_2 +_n v_2 \times_n \partial v_2/\partial x_3) -_n$$

$$-_n \partial/\partial x_3(-v_1 \times_n \partial v_2/\partial x_1 +_n v_1 \times_n \partial v_1/\partial x_2 +_n v_3 \times_n \partial v_3/\partial x_2 -_n v_3 \times_n \partial v_2/\partial x_3) +_n \omega_{42}^1 =$$

$$= -\partial v_1/\partial x_2 \times_n \partial v_3/\partial x_1 -_n v_1 \times_n \partial/\partial x_2(\partial v_3/\partial x_1) +_n \partial v_1/\partial x_2 \times_n \partial v_1/\partial x_3 +_n$$

$$+_n v_1 \times_n \partial/\partial x_2(\partial v_1/\partial x_3) -_n$$

$$-_n \partial v_2/\partial x_2 \times_n \partial v_3/\partial x_2 -_n v_2 \times_n \partial/\partial x_2(\partial v_3/\partial x_2) +_n \partial v_2/\partial x_2 \times_n \partial v_2/\partial x_3 +_n$$

$$+_n v_2 \times_n \partial/\partial x_2(\partial v_2/\partial x_3) +_n$$

$$+_n \partial v_1/\partial x_3 \times_n \partial v_2/\partial x_1 +_n v_1 \times_n \partial/\partial x_3(\partial v_2/\partial x_1) -_n \partial v_1/\partial x_3 \times_n \partial v_1/\partial x_2 -_n$$

$$-_n v_1 \times_n \partial/\partial x_3(\partial v_1/\partial x_2) -_n$$

$$-_n \partial v_3/\partial x_3 \times_n \partial v_3/\partial x_2 -_n v_3 \times_n \partial/\partial x_3(\partial v_3/\partial x_2) +_n \partial v_3/\partial x_3 \times_n \partial v_2/\partial x_3 +_n$$

$$+_n v_3 \times_n \partial/\partial x_3 (\partial v_2/\partial x_3) +_n$$

$$+_n \omega_{42}^1 +_n \omega_{43}^1 =$$

$$= -\partial v_1/\partial x_2 \times_n \partial v_3/\partial x_1 -_n v_1 \times_n \partial^2 v_3/\partial x_1 \partial x_2 +_n \partial v_1/\partial x_2 \times_n \partial v_1/\partial x_3 +_n$$

$$+_n v_1 \times_n \partial^2 v_1/\partial x_2 \partial x_3 -_n$$

$$-_n \partial v_2/\partial x_2 \times_n \partial v_3/\partial x_2 -_n v_2 \times_n \partial^2 v_3/\partial x_2{}^2 +_n \partial v_2/\partial x_2 \times_n \partial v_2/\partial x_3 +_n$$

$$+_n v_2 \times_n \partial^2 v_2/\partial x_2 \partial x_3 +_n$$

$$+_n \partial v_1/\partial x_3 \times_n \partial v_2/\partial x_1 +_n v_1 \times_n \partial^2 v_2/\partial x_1 \partial x_3 -_n \partial v_1/\partial x_3 \times_n \partial v_1/\partial x_2 -_n$$
$$-_n v_1 \times_n \partial^2 v_1/\partial x_2 \partial x_3 -_n$$

$$-_n \partial v_3/\partial x_3 \times_n \partial v_3/\partial x_2 -_n v_3 \times_n \partial^2 v_3/\partial x_2 \partial x_3 +_n \partial v_3/\partial x_3 \times_n \partial v_2/\partial x_3 +_n$$
$$+_n v_3 \times_n \partial^2 v_2/\partial x_3{}^2 +_n$$

$$+_n \omega_{42}^1 +_n \omega_{43}^1 +_n \omega_{44}^1$$

where $\omega_{42}^1, \omega_{43}^1, \omega_{44}^1$ are the random variables depend on n and m.

$$R_2 = -\partial/\partial x_1(-v_1 \times_n \partial v_3/\partial x_1 +_n v_1 \times_n \partial v_1/\partial x_3 -_n v_2 \times_n \partial v_3/\partial x_2 +_n v_2$$
$$\times_n \partial v_2/\partial x_3) +_n$$

$$+_n \partial/\partial x_3(v_2 \times_n \partial v_2/\partial x_1 -_n v_2 \times_n \partial v_1/\partial x_2 +_n v_3 \times_n \partial v_3/\partial x_1 -_n v_3$$
$$\times_n \partial v_1/\partial x_3) +_n \omega_{42}^2 =$$

$$= \partial v_1/\partial x_1 \times_n \partial v_3/\partial x_1 +_n v_1 \times_n \partial/\partial x_1(\partial v_3/\partial x_1) -_n \partial v_1/\partial x_1 \times_n \partial v_1/\partial x_3 -_n$$

$$-_n v_1 \times_n \partial/\partial x_1(\partial v_1/\partial x_3) +_n$$

$$+_n \partial v_2/\partial x_1 \times_n \partial v_3/\partial x_2 +_n v_2 \times_n \partial/\partial x_1(\partial v_3/\partial x_2) -_n \partial v_2/\partial x_1 \times_n \partial v_2/\partial x_3 -_n$$

$$-_n v_2 \times_n \partial/\partial x_1(\partial v_2/\partial x_3) +_n$$

$$+_n \partial v_2/\partial x_3 \times_n \partial v_2/\partial x_1 +_n v_2 \times_n \partial/\partial x_3(\partial v_2/\partial x_1) -_n \partial v_2/\partial x_3 \times_n \partial v_1/\partial x_2 -_n$$

$$-_n v_2 \times_n \partial/\partial x_3(\partial v_1/\partial x_2) +_n$$

$$+_n \partial v_3/\partial x_3 \times_n \partial v_3/\partial x_1 +_n v_3 \times_n \partial/\partial x_3(\partial v_3/\partial x_1) -_n \partial v_3/\partial x_3 \times_n \partial v_1/\partial x_3 -_n$$

$$-_n v_3 \times_n \partial/\partial x_3(\partial v_1/\partial x_3) +_n$$

$$+_n \omega_{42}^2 +_n \omega_{43}^2 =$$

$$= \partial v_1/\partial x_1 \times_n \partial v_3/\partial x_1 +_n v_1 \times_n \partial^2 v_3/\partial x_1^2 -_n \partial v_1/\partial x_1 \times_n \partial v_1/\partial x_3 -_n$$

$$-_n v_1 \times_n \partial^2 v_1/\partial x_1 \partial x_3 +_n$$

$$+_n \partial v_2/\partial x_1 \times_n \partial v_3/\partial x_2 +_n v_2 \times_n \partial^2 v_3/\partial x_1 \partial x_2 -_n \partial v_2/\partial x_1 \times_n \partial v_2/\partial x_3 -_n$$

$$-_n v_2 \times_n \partial^2 v_2/\partial x_1 \partial x_3 +_n$$

$$+_n \partial v_2/\partial x_3 \times_n \partial v_2/\partial x_1 +_n v_2 \times_n \partial^2 v_2/\partial x_1 \partial x_3 -_n \partial v_2/\partial x_3 \times_n \partial v_1/\partial x_2 -_n$$

$$-_n v_2 \times_n \partial^2 v_1/\partial x_2 \partial x_3 +_n$$

$$+_n \partial v_3/\partial x_3 \times_n \partial v_3/\partial x_1 +_n v_3 \times_n \partial^2 v_3/\partial x_1 \partial x_3 -_n \partial v_3/\partial x_3 \times_n \partial v_1/\partial x_3 -_n$$

$$-_n v_3 \times_n \partial^2 v_1/\partial x_3^2 +_n$$

$$+_n\omega_{42}^2 +_n \omega_{43}^2 +_n \omega_{44}^2$$

where $\omega_{42}^2, \omega_{43}^2, \omega_{44}^2$ are the random variables depend on n and m.

$$R_3 = \partial/\partial x_1(-v_1 \times_n \partial v_2/\partial x_1 +_n v_1 \times_n \partial v_1/\partial x_2 +_n v_3 \times_n \partial v_3/\partial x_2$$
$$-_n v_3 \times_n \partial v_2/\partial x_3)-_n$$

$$-_n \partial/\partial x_2(v_2 \times_n \partial v_2/\partial x_1 -_n v_2 \times_n \partial v_1/\partial x_2 +_n v_3 \times_n \partial v_3/\partial x_1$$
$$-_n v_3 \times_n \partial v_1/\partial x_3) +_n \omega_{42}^3 =$$

$$= -\partial v_1/\partial x_1 \times_n \partial v_2/\partial x_1 -_n v_1 \times_n \partial/\partial x_1(\partial v_2/\partial x_1) +_n \partial v_1/\partial x_1 \times_n \partial v_1/\partial x_2 +_n$$

$$+_n v_1 \times_n \partial/\partial x_1(\partial v_1/\partial x_2) +_n$$

$$+_n \partial v_3/\partial x_1 \times_n \partial v_3/\partial x_2 +_n v_3 \times_n \partial/\partial x_1(\partial v_3/\partial x_2) -_n \partial v_3/\partial x_1 \times_n \partial v_2/\partial x_3 -_n$$

$$-_n v_3 \times_n \partial/\partial x_1(\partial v_2/\partial x_3) -_n$$

$$-_n \partial v_2/\partial x_2 \times_n \partial v_2/\partial x_1 -_n v_2 \times_n \partial/\partial x_2(\partial v_2/\partial x_1) +_n \partial v_2/\partial x_2 \times_n \partial v_1/\partial x_2 +_n$$

$$+_n v_2 \times_n \partial/\partial x_2(\partial v_1/\partial x_2) -_n$$

$$-_n \partial v_3/\partial x_2 \times_n \partial v_3/\partial x_1 -_n v_3 \times_n \partial/\partial x_2(\partial v_3/\partial x_1) +_n \partial v_3/\partial x_2 \times_n \partial v_1/\partial x_3 +_n$$

$$+_n v_3 \times_n \partial/\partial x_2(\partial v_1/\partial x_3) +_n$$

$$+_n \omega_{42}^3 +_n \omega_{43}^3 =$$

$$= -\partial v_1/\partial x_1 \times_n \partial v_2/\partial x_1 -_n v_1 \times_n \partial^2 v_2/\partial x_1^2 +_n \partial v_1/\partial x_1 \times_n \partial v_1/\partial x_2 +_n$$

$$+_n v_1 \times_n \partial^2 v_1/\partial x_1 \partial x_2 +_n$$

$+_n \partial v_3/\partial x_1 \times_n \partial v_3/\partial x_2 +_n v_3 \times_n \partial^2 v_3/\partial x_1 \partial x_2 -_n \partial v_3/\partial x_1 \times_n \partial v_2/\partial x_3 -_n$

$$-_n v_3 \times_n \partial^2 v_2/\partial x_1 \partial x_3 -_n$$

$-_n \partial v_2/\partial x_2 \times_n \partial v_2/\partial x_1 -_n v_2 \times_n \partial^2 v_2/\partial x_1 \partial x_2 +_n \partial v_2/\partial x_2 \times_n \partial v_1/\partial x_2 +_n$

$$+_n v_2 \times_n \partial^2 v_1/\partial x_2{}^2 -_n$$

$-_n \partial v_3/\partial x_2 \times_n \partial v_3/\partial x_1 -_n v_3 \times_n \partial^2 v_3/\partial x_1 \partial x_2 +_n \partial v_3/\partial x_2 \times_n \partial v_1/\partial x_3 +_n$

$$+_n v_3 \times_n \partial^2 v_1/\partial x_2 \partial x_3 +_n$$

$$+_n \omega_{42}^3 +_n \omega_{43}^3 +_n \omega_{44}^3$$

where $\omega_{42}^3, \omega_{43}^3, \omega_{44}^3$ are the random variables depend on n and m. Also we have

$$\mathbf{rot}\ \mathbf{g} = (\partial g_3/\partial x_2 -_n \partial g_2/\partial x_3, \partial g_1/\partial x_3 -_n \partial g_3/\partial x_1, \partial g_2/\partial x_1 -_n \partial g_1/\partial x_2)$$

If we generalize gravity to any given externally applied force $\mathbf{f}$, we have to change term

$$\rho \times_n \mathbf{rot}\ \mathbf{g}$$

to

$$\mathbf{rot}\ \mathbf{f} = (\partial f_3/\partial x_2 -_n \partial f_2/\partial x_3, \partial f_1/\partial x_3 -_n \partial f_3/\partial x_1, \partial f_2/\partial x_1 -_n \partial f_1/\partial x_2)$$

We have

$$\Delta \mathbf{v} = (\partial^2 v_1/\partial x_1{}^2 +_n \partial^2 v_1/\partial x_2{}^2 +_n \partial^2 v_1/\partial x_3{}^2,$$

$$\partial^2 v_2/\partial x_1{}^2 +_n \partial^2 v_2/\partial x_2{}^2 +_n \partial^2 v_2/\partial x_3{}^2,$$

$$\partial^2 v_3/\partial x_1{}^2 +_n \partial^2 v_3/\partial x_2{}^2 +_n \partial^2 v_3/\partial x_3{}^2)$$

We get

$$\mathbf{rot}\ \Delta \mathbf{v} = (\partial(\partial^2 v_3/\partial x_1{}^2 +_n \partial^2 v_3/\partial x_2{}^2 +_n \partial^2 v_3/\partial x_3{}^2))/\partial x_2 -_n$$

$$-_n \partial(\partial^2 v_2/\partial x_1{}^2 +_n \partial^2 v_2/\partial x_2{}^2 +_n \partial^2 v_2/\partial x_3{}^2)/\partial x_3,$$

$$\partial(\partial^2 v_1/\partial x_1{}^2 +_n \partial^2 v_1/\partial x_2{}^2 +_n \partial^2 v_1/\partial x_3{}^2)/\partial x_3 -_n$$

$$-_n \partial(\partial^2 v_3/\partial x_1{}^2 +_n \partial^2 v_3/\partial x_2{}^2 +_n \partial^2 v_3/\partial x_3{}^2)/\partial x_1,$$

$$\partial(\partial^2 v_2/\partial x_1{}^2 +_n \partial^2 v_2/\partial x_2{}^2 +_n \partial^2 v_2/\partial x_3{}^2)/\partial x_1 -_n$$

$$-_n \partial(\partial^2 v_1/\partial x_1{}^2 +_n \partial^2 v_1/\partial x_2{}^2 +_n \partial^2 v_1/\partial x_3{}^2)/\partial x_2) =$$

$$= (Q_1, Q_2, Q_3) +_n (\omega_{45}^1, \omega_{45}^2, \omega_{45}^3)$$

where

$$\omega_{45} = (\omega_{45}^1, \omega_{45}^2, \omega_{45}^3)$$

is the random vector with random coordinates depend on n and m. And

$$Q_1 = \partial^3 v_3/\partial x_1{}^2 \partial x_2 +_n \partial^3 v_3/\partial x_2{}^3 +_n \partial^3 v_3/\partial x_2 \partial x_3{}^2 -_n \partial^3 v_2/\partial x_1{}^2 \partial x_3 -_n$$

$$-_n \partial^3 v_2/\partial x_2{}^2 \partial x_3 -_n \partial^3 v_2/\partial x_3{}^3$$

$$Q_2 = \partial^3 v_1/\partial x_1{}^2 \partial x_3 +_n \partial^3 v_1/\partial x_2{}^2 \partial x_3 +_n \partial^3 v_1/\partial x_3{}^3 -_n \partial^3 v_3/\partial x_1{}^3 -_n$$

$$-_n \partial^3 v_3/\partial x_1 \partial x_2{}^2 -_n \partial^3 v_3/\partial x_1 \partial x_3{}^2$$

$$Q_3 = \partial^3 v_2/\partial x_1{}^3 +_n \partial^3 v_2/\partial x_1 \partial x_2{}^2 +_n \partial^3 v_2/\partial x_1 \partial x_3{}^2 -_n \partial^3 v_1/\partial x_1{}^2 \partial x_2 -_n$$

$$-_n \partial^3 v_1/\partial x_2{}^3 -_n \partial^3 v_1/\partial x_2 \partial x_3{}^2$$

We rewrite now the first equation of the system

$$\begin{cases} \rho \times_n (\partial/\partial t(\mathbf{rot\ v}) -_n \mathbf{rot\ (v} \times \mathbf{rot\ v})) = \mathbf{rot\ f} +_n \alpha \times_n \mathbf{rot} \Delta \mathbf{v} +_n \omega_{38} \\ div\mathbf{v} = \omega_{28} \\ \alpha > 0 \end{cases}$$

in coordinates.

By coordinate x_1 we have

$$\rho \times_n (\partial^2 v_3/\partial x_2 \partial t -_n \partial^2 v_2/\partial x_3 \partial t +_n \omega_{39}^1 +_n \omega_{39}^4 -_n$$

$$-_n(-\partial v_1/\partial x_2 \times_n \partial v_3/\partial x_1 -_n v_1 \times_n \partial^2 v_3/\partial x_1 \partial x_2 +_n \partial v_1/\partial x_2 \times_n \partial v_1/\partial x_3 +_n$$

$$+_n v_1 \times_n \partial^2 v_1/\partial x_2 \partial x_3 -_n$$

$$-_n \partial v_2/\partial x_2 \times_n \partial v_3/\partial x_2 -_n v_2 \times_n \partial^2 v_3/\partial x_2{}^2 +_n \partial v_2/\partial x_2 \times_n \partial v_2/\partial x_3 +_n$$

$$+_n v_2 \times_n \partial^2 v_2/\partial x_2 \partial x_3 +_n$$

$$+_n \partial v_1/\partial x_3 \times_n \partial v_2/\partial x_1 +_n v_1 \times_n \partial^2 v_2/\partial x_1 \partial x_3 -_n \partial v_1/\partial x_3 \times_n \partial v_1/\partial x_2 -_n$$

$$-_n v_1 \times_n \partial^2 v_1/\partial x_2 \partial x_3 -_n$$

$$-_n \partial v_3/\partial x_3 \times_n \partial v_3/\partial x_2 -_n v_3 \times_n \partial^2 v_3/\partial x_2 \partial x_3 +_n \partial v_3/\partial x_3 \times_n \partial v_2/\partial x_3 +_n$$

$$+_n v_3 \times_n \partial^2 v_2/\partial x_3{}^2 +_n$$

$$+_n \omega_{41}^1 +_n \omega_{42}^1 +_n \omega_{43}^1 +_n \omega_{44}^1)) =$$

$$= \partial f_3/\partial x_2 -_n \partial f_2/\partial x_3 +_n$$

$$+_n \alpha \times_n (\partial^3 v_3/\partial x_1{}^2 \partial x_2 +_n \partial^3 v_3/\partial x_2{}^3 +_n \partial^3 v_3/\partial x_2 \partial x_3{}^2 -_n \partial^3 v_2/\partial x_1{}^2 \partial x_3 -_n$$

$$-_n \partial^3 v_2/\partial x_2{}^2 \partial x_3 -_n \partial^3 v_2/\partial x_3{}^3 +_n \omega_{45}^1) +_n \omega_{38}^1$$

By coordinate x_2 we have

$$\rho \times_n (\partial^2 v_1/\partial x_3 \partial t -_n \partial^2 v_3/\partial x_1 \partial t +_n \omega_{39}^2 +_n \omega_{39}^5 -_n$$

$$-_n(\partial v_1/\partial x_1 \times_n \partial v_3/\partial x_1 +_n v_1 \times_n \partial^2 v_3/\partial x_1{}^2 -_n \partial v_1/\partial x_1 \times_n \partial v_1/\partial x_3 -_n$$

$$-_n v_1 \times_n \partial^2 v_1/\partial x_1 \partial x_3 +_n$$

$$+_n \partial v_2/\partial x_1 \times_n \partial v_3/\partial x_2 +_n v_2 \times_n \partial^2 v_3/\partial x_1 \partial x_2 -_n \partial v_2/\partial x_1 \times_n \partial v_2/\partial x_3 -_n$$

$$-_n v_2 \times_n \partial^2 v_2/\partial x_1 \partial x_3 +_n$$

$$+_n \partial v_2/\partial x_3 \times_n \partial v_2/\partial x_1 +_n v_2 \times_n \partial^2 v_2/\partial x_1 \partial x_3 -_n \partial v_2/\partial x_3 \times_n \partial v_1/\partial x_2 -_n$$

$$-_n v_2 \times_n \partial^2 v_1/\partial x_2 \partial x_3 +_n$$

$$+_n \partial v_3/\partial x_3 \times_n \partial v_3/\partial x_1 +_n v_3 \times_n \partial^2 v_3/\partial x_1 \partial x_3 -_n \partial v_3/\partial x_3 \times_n \partial v_1/\partial x_3 -_n$$

$$-_n v_3 \times_n \partial^2 v_1/\partial x_3{}^2 +_n$$

$$+_n \omega_{41}^2 +_n \omega_{42}^2 +_n \omega_{43}^2 +_n \omega_{44}^2)) =$$

$$= \partial f_1/\partial x_3 -_n \partial f_3/\partial x_1 +_n$$

$$+_n \alpha \times_n (\partial^3 v_1/\partial x_1{}^2 \partial x_3 +_n \partial^3 v_1/\partial x_2{}^2 \partial x_3 +_n \partial^3 v_1/\partial x_3{}^3 -_n \partial^3 v_3/\partial x_1{}^3 -_n$$

$$-_n \partial^3 v_3/\partial x_1 \partial x_2{}^2 -_n \partial^3 v_3/\partial x_1 \partial x_3{}^2 +_n \omega_{45}^2) +_n \omega_{38}^2$$

By coordinate x_3 we have

$$\rho \times_n (\partial^2 v_2/\partial x_1 \partial t -_n \partial^2 v_1/\partial x_2 \partial t +_n \omega_{39}^3 +_n \omega_{39}^6 -_n$$

$$-_n (-\partial v_1/\partial x_1 \times_n \partial v_2/\partial x_1 -_n v_1 \times_n \partial^2 v_2/\partial x_1{}^2 +_n \partial v_1/\partial x_1 \times_n \partial v_1/\partial x_2 +_n$$

$$+_n v_1 \times_n \partial^2 v_1/\partial x_1 \partial x_2 +_n$$

$$+_n \partial v_3/\partial x_1 \times_n \partial v_3/\partial x_2 +_n v_3 \times_n \partial^2 v_3/\partial x_1 \partial x_2 -_n \partial v_3/\partial x_1 \times_n \partial v_2/\partial x_3 -_n$$

$$-_n v_3 \times_n \partial^2 v_2/\partial x_1 \partial x_3 -_n$$

$$-_n \partial v_2/\partial x_2 \times_n \partial v_2/\partial x_1 -_n v_2 \times_n \partial^2 v_2/\partial x_1 \partial x_2 +_n \partial v_2/\partial x_2 \times_n \partial v_1/\partial x_2 +_n$$

$$+_n v_2 \times_n \partial^2 v_1/\partial x_2{}^2 -_n$$

$$-_n \partial v_3/\partial x_2 \times_n \partial v_3/\partial x_1 -_n v_3 \times_n \partial^2 v_3/\partial x_1 \partial x_2 +_n \partial v_3/\partial x_2 \times_n \partial v_1/\partial x_3 +_n$$

$$+_n v_3 \times_n \partial^2 v_1/\partial x_2 \partial x_3 +_n$$

$$+_n \omega_{41}^3 +_n \omega_{42}^3 +_n \omega_{43}^3 +_n \omega_{44}^3)) =$$

$$= \partial f_2/\partial x_1 -_n \partial f_1/\partial x_2 +_n$$

$$+_n \alpha \times_n (\partial^3 v_2/\partial x_1{}^3 +_n \partial^3 v_2/\partial x_1 \partial x_2{}^2 +_n \partial^3 v_2/\partial x_1 \partial x_3{}^2 -_n \partial^3 v_1/\partial x_1{}^2 \partial x_2 -_n$$

$$-_n \partial^3 v_1/\partial x_2{}^3 -_n \partial^3 v_1/\partial x_2 \partial x_3{}^2 +_n \omega_{45}^3) +_n \omega_{38}^3$$

So, probability of the first equation of Navier-Stokes system (in coordinates) in Mathematics of Observers can be written as:

by coordinate x_1

$$\rho \times_n (\partial^2 v_3/\partial x_2 \partial t -_n \partial^2 v_2/\partial x_3 \partial t -_n$$

$$-_n(-\partial v_1/\partial x_2 \times_n \partial v_3/\partial x_1 -_n v_1 \times_n \partial^2 v_3/\partial x_1 \partial x_2 +_n \partial v_1/\partial x_2 \times_n \partial v_1/\partial x_3 +_n$$

$$+_n v_1 \times_n \partial^2 v_1/\partial x_2 \partial x_3 -_n$$

$$-_n \partial v_2/\partial x_2 \times_n \partial v_3/\partial x_2 -_n v_2 \times_n \partial^2 v_3/\partial x_2{}^2 +_n \partial v_2/\partial x_2 \times_n \partial v_2/\partial x_3 +_n$$

$$+_n v_2 \times_n \partial^2 v_2/\partial x_2 \partial x_3 +_n$$

$$+_n \partial v_1/\partial x_3 \times_n \partial v_2/\partial x_1 +_n v_1 \times_n \partial^2 v_2/\partial x_1 \partial x_3 -_n \partial v_1/\partial x_3 \times_n \partial v_1/\partial x_2 -_n$$

$$-_n v_1 \times_n \partial^2 v_1/\partial x_2 \partial x_3 -_n$$

$$-_n \partial v_3/\partial x_3 \times_n \partial v_3/\partial x_2 -_n v_3 \times_n \partial^2 v_3/\partial x_2 \partial x_3 +_n \partial v_3/\partial x_3 \times_n \partial v_2/\partial x_3 +_n$$

$$+_n v_3 \times_n \partial^2 v_2/\partial x_3^2)) =$$

$$= \partial f_3/\partial x_2 -_n \partial f_2/\partial x_3 +_n$$

$$+_n \alpha \times_n (\partial^3 v_3/\partial x_1^2 \partial x_2 +_n \partial^3 v_3/\partial x_2^3 +_n \partial^3 v_3/\partial x_2 \partial x_3^2 -_n \partial^3 v_2/\partial x_1^2 \partial x_3 -_n$$

$$-_n \partial^3 v_2/\partial x_2^2 \partial x_3 -_n \partial^3 v_2/\partial x_3^3)$$

by coordinate x_2

$$\rho \times_n (\partial^2 v_1/\partial x_3 \partial t -_n \partial^2 v_3/\partial x_1 \partial t -_n$$

$$-_n (\partial v_1/\partial x_1 \times_n \partial v_3/\partial x_1 +_n v_1 \times_n \partial^2 v_3/\partial x_1^2 -_n \partial v_1/\partial x_1 \times_n \partial v_1/\partial x_3 -_n$$

$$-_n v_1 \times_n \partial^2 v_1/\partial x_1 \partial x_3 +_n$$

$$+_n \partial v_2/\partial x_1 \times_n \partial v_3/\partial x_2 +_n v_2 \times_n \partial^2 v_3/\partial x_1 \partial x_2 -_n \partial v_2/\partial x_1 \times_n \partial v_2/\partial x_3 -_n$$

$$-_n v_2 \times_n \partial^2 v_2/\partial x_1 \partial x_3 +_n$$

$$+_n \partial v_2/\partial x_3 \times_n \partial v_2/\partial x_1 +_n v_2 \times_n \partial^2 v_2/\partial x_1 \partial x_3 -_n \partial v_2/\partial x_3 \times_n \partial v_1/\partial x_2 -_n$$

$$-_n v_2 \times_n \partial^2 v_1/\partial x_2 \partial x_3 +_n$$

$$+_n \partial v_3/\partial x_3 \times_n \partial v_3/\partial x_1 +_n v_3 \times_n \partial^2 v_3/\partial x_1 \partial x_3 -_n \partial v_3/\partial x_3 \times_n \partial v_1/\partial x_3 -_n$$

$$-_n v_3 \times_n \partial^2 v_1/\partial x_3^2)) =$$

$$= \partial f_1/\partial x_3 -_n \partial f_3/\partial x_1 +_n$$

$$+_n \alpha \times_n (\partial^3 v_1/\partial x_1^2 \partial x_3 +_n \partial^3 v_1/\partial x_2^2 \partial x_3 +_n \partial^3 v_1/\partial x_3^3 -_n \partial^3 v_3/\partial x_1^3 -_n$$

$$-_n \partial^3 v_3/\partial x_1 \partial x_2^2 -_n \partial^3 v_3/\partial x_1 \partial x_3^2)$$

by coordinate x_3

$$\rho \times_n (\partial^2 v_2/\partial x_1 \partial t -_n \partial^2 v_1/\partial x_2 \partial t -_n$$

$$-_n(-\partial v_1/\partial x_1 \times_n \partial v_2/\partial x_1 -_n v_1 \times_n \partial^2 v_2/\partial x_1{}^2 +_n \partial v_1/\partial x_1 \times_n \partial v_1/\partial x_2 +_n$$

$$+_n v_1 \times_n \partial^2 v_1/\partial x_1 \partial x_2 +_n$$

$$+_n \partial v_3/\partial x_1 \times_n \partial v_3/\partial x_2 +_n v_3 \times_n \partial^2 v_3/\partial x_1 \partial x_2 -_n \partial v_3/\partial x_1 \times_n \partial v_2/\partial x_3 -_n$$

$$-_n v_3 \times_n \partial^2 v_2/\partial x_1 \partial x_3 -_n$$

$$-_n \partial v_2/\partial x_2 \times_n \partial v_2/\partial x_1 -_n v_2 \times_n \partial^2 v_2/\partial x_1 \partial x_2 +_n \partial v_2/\partial x_2 \times_n \partial v_1/\partial x_2 +_n$$

$$+_n v_2 \times_n \partial^2 v_1/\partial x_2{}^2 -_n$$

$$-_n \partial v_3/\partial x_2 \times_n \partial v_3/\partial x_1 -_n v_3 \times_n \partial^2 v_3/\partial x_1 \partial x_2 +_n \partial v_3/\partial x_2 \times_n \partial v_1/\partial x_3 +_n$$

$$+_n v_3 \times_n \partial^2 v_1/\partial x_2 \partial x_3)) =$$

$$= \partial f_2/\partial x_1 -_n \partial f_1/\partial x_2 +_n$$

$$+_n \alpha \times_n (\partial^3 v_2/\partial x_1{}^3 +_n \partial^3 v_2/\partial x_1 \partial x_2{}^2 +_n \partial^3 v_2/\partial x_1 \partial x_3{}^2 -_n \partial^3 v_1/\partial x_1{}^2 \partial x_2 -_n$$

$$-_n \partial^3 v_1/\partial x_2{}^3 -_n \partial^3 v_1/\partial x_2 \partial x_3{}^2)$$

is less than 1.
So, we proved

Theorem 11.29 *The first equation of Navier-Stokes system (in coordinates) in Mathematics of Observers can be written as:*
  by coordinate x_1

$$\rho \times_n (\partial^2 v_3/\partial x_2 \partial t -_n \partial^2 v_2/\partial x_3 \partial t +_n \omega_{39}^1 +_n \omega_{39}^4 -_n$$

$$-_n(-\partial v_1/\partial x_2 \times_n \partial v_3/\partial x_1 -_n v_1 \times_n \partial^2 v_3/\partial x_1 \partial x_2 +_n \partial v_1/\partial x_2 \times_n \partial v_1/\partial x_3 +_n$$

$$+_n v_1 \times_n \partial^2 v_1/\partial x_2 \partial x_3 -_n$$

$$-_n \partial v_2/\partial x_2 \times_n \partial v_3/\partial x_2 -_n v_2 \times_n \partial^2 v_3/\partial x_2{}^2 +_n \partial v_2/\partial x_2 \times_n \partial v_2/\partial x_3 +_n$$

$$+_n v_2 \times_n \partial^2 v_2/\partial x_2 \partial x_3 +_n$$

$$+_n \partial v_1/\partial x_3 \times_n \partial v_2/\partial x_1 +_n v_1 \times_n \partial^2 v_2/\partial x_1 \partial x_3 -_n \partial v_1/\partial x_3 \times_n \partial v_1/\partial x_2 -_n$$

$$-_n v_1 \times_n \partial^2 v_1/\partial x_2 \partial x_3 -_n$$

$$-_n \partial v_3/\partial x_3 \times_n \partial v_3/\partial x_2 -_n v_3 \times_n \partial^2 v_3/\partial x_2 \partial x_3 +_n \partial v_3/\partial x_3 \times_n \partial v_2/\partial x_3 +_n$$

$$+_n v_3 \times_n \partial^2 v_2/\partial x_3{}^2 +_n$$

$$+_n \omega_{41}^1 +_n \omega_{42}^1 +_n \omega_{43}^1 +_n \omega_{44}^1)) =$$

$$= \partial f_3/\partial x_2 -_n \partial f_2/\partial x_3 +_n$$

$$+_n \alpha \times_n (\partial^3 v_3/\partial x_1{}^2 \partial x_2 +_n \partial^3 v_3/\partial x_2{}^3 +_n \partial^3 v_3/\partial x_2 \partial x_3{}^2 -_n \partial^3 v_2/\partial x_1{}^2 \partial x_3 -_n$$

$$-_n \partial^3 v_2/\partial x_2{}^2 \partial x_3 -_n \partial^3 v_2/\partial x_3{}^3 +_n \omega_{45}^1) +_n \omega_{38}^1$$

by coordinate x_2

$$\rho \times_n (\partial^2 v_1/\partial x_3 \partial t -_n \partial^2 v_3/\partial x_1 \partial t +_n \omega_{39}^2 +_n \omega_{39}^5 -_n$$

$$-_n (\partial v_1/\partial x_1 \times_n \partial v_3/\partial x_1 +_n v_1 \times_n \partial^2 v_3/\partial x_1{}^2 -_n \partial v_1/\partial x_1 \times_n \partial v_1/\partial x_3 -_n$$

$$-_n v_1 \times_n \partial^2 v_1/\partial x_1 \partial x_3 +_n$$

$$+_n \partial v_2/\partial x_1 \times_n \partial v_3/\partial x_2 +_n v_2 \times_n \partial^2 v_3/\partial x_1 \partial x_2 -_n \partial v_2/\partial x_1 \times_n \partial v_2/\partial x_3 -_n$$

$$-_n v_2 \times_n \partial^2 v_2/\partial x_1 \partial x_3 +_n$$

$$+_n \partial v_2/\partial x_3 \times_n \partial v_2/\partial x_1 +_n v_2 \times_n \partial^2 v_2/\partial x_1 \partial x_3 -_n \partial v_2/\partial x_3 \times_n \partial v_1/\partial x_2 -_n$$

$$-_n v_2 \times_n \partial^2 v_1/\partial x_2 \partial x_3 +_n$$

$$+_n \partial v_3/\partial x_3 \times_n \partial v_3/\partial x_1 +_n v_3 \times_n \partial^2 v_3/\partial x_1 \partial x_3 -_n \partial v_3/\partial x_3 \times_n \partial v_1/\partial x_3 -_n$$

$$-_n v_3 \times_n \partial^2 v_1/\partial x_3{}^2 +_n$$

$$+_n \omega_{41}^2 +_n \omega_{42}^2 +_n \omega_{43}^2 +_n \omega_{44}^2)) =$$

$$= \partial f_1/\partial x_3 -_n \partial f_3/\partial x_1 +_n$$

$$+_n \alpha \times_n (\partial^3 v_1/\partial x_1{}^2 \partial x_3 +_n \partial^3 v_1/\partial x_2{}^2 \partial x_3 +_n \partial^3 v_1/\partial x_3{}^3 -_n \partial^3 v_3/\partial x_1{}^3 -_n$$

$$-_n \partial^3 v_3/\partial x_1 \partial x_2{}^2 -_n \partial^3 v_3/\partial x_1 \partial x_3{}^2 +_n \omega_{45}^2) +_n \omega_{38}^2$$

by coordinate x_3

$$\rho \times_n (\partial^2 v_2/\partial x_1 \partial t -_n \partial^2 v_1/\partial x_2 \partial t +_n \omega_{39}^3 +_n \omega_{39}^6 -_n$$

$$-_n (-\partial v_1/\partial x_1 \times_n \partial v_2/\partial x_1 -_n v_1 \times_n \partial^2 v_2/\partial x_1{}^2 +_n \partial v_1/\partial x_1 \times_n \partial v_1/\partial x_2 +_n$$

$$+_n v_1 \times_n \partial^2 v_1/\partial x_1 \partial x_2 +_n$$

$$+_n \partial v_3/\partial x_1 \times_n \partial v_3/\partial x_2 +_n v_3 \times_n \partial^2 v_3/\partial x_1 \partial x_2 -_n \partial v_3/\partial x_1 \times_n \partial v_2/\partial x_3 -_n$$

$$-_n v_3 \times_n \partial^2 v_2/\partial x_1 \partial x_3 -_n$$

$$-_n \partial v_2/\partial x_2 \times_n \partial v_2/\partial x_1 -_n v_2 \times_n \partial^2 v_2/\partial x_1 \partial x_2 +_n \partial v_2/\partial x_2 \times_n \partial v_1/\partial x_2 +_n$$

$$+_n v_2 \times_n \partial^2 v_1/\partial x_2^2 -_n$$

$$-_n \partial v_3/\partial x_2 \times_n \partial v_3/\partial x_1 -_n v_3 \times_n \partial^2 v_3/\partial x_1 \partial x_2 +_n \partial v_3/\partial x_2 \times_n \partial v_1/\partial x_3 +_n$$

$$+_n v_3 \times_n \partial^2 v_1/\partial x_2 \partial x_3 +_n$$

$$+_n \omega_{41}^3 +_n \omega_{42}^3 +_n \omega_{43}^3 +_n \omega_{44}^3)) =$$

$$= \partial f_2/\partial x_1 -_n \partial f_1/\partial x_2 +_n$$

$$+_n \alpha \times_n (\partial^3 v_2/\partial x_1^3 +_n \partial^3 v_2/\partial x_1 \partial x_2^2 +_n \partial^3 v_2/\partial x_1 \partial x_3^2 -_n \partial^3 v_1/\partial x_1^2 \partial x_2 -_n$$

$$-_n \partial^3 v_1/\partial x_2^3 -_n \partial^3 v_1/\partial x_2 \partial x_3^2 +_n \omega_{45}^3) +_n \omega_{38}^3$$

We assume that all elements of these equalities belong to W_n.

Corollary 11.30 *The probability of the first equation of Navier-Stokes system (in coordinates) in Mathematics of Observers can be written as:*
   by coordinate x_1

$$\rho \times_n (\partial^2 v_3/\partial x_2 \partial t -_n \partial^2 v_2/\partial x_3 \partial t -_n$$

$$-_n(-\partial v_1/\partial x_2 \times_n \partial v_3/\partial x_1 -_n v_1 \times_n \partial^2 v_3/\partial x_1 \partial x_2 +_n \partial v_1/\partial x_2 \times_n \partial v_1/\partial x_3 +_n$$

$$+_n v_1 \times_n \partial^2 v_1/\partial x_2 \partial x_3 -_n$$

$$-_n \partial v_2/\partial x_2 \times_n \partial v_3/\partial x_2 -_n v_2 \times_n \partial^2 v_3/\partial x_2^2 +_n \partial v_2/\partial x_2 \times_n \partial v_2/\partial x_3 +_n$$

$$+_n v_2 \times_n \partial^2 v_2/\partial x_2 \partial x_3 +_n$$

$$+_n \partial v_1/\partial x_3 \times_n \partial v_2/\partial x_1 +_n v_1 \times_n \partial^2 v_2/\partial x_1 \partial x_3 -_n \partial v_1/\partial x_3 \times_n \partial v_1/\partial x_2 -_n$$

$$-_n v_1 \times_n \partial^2 v_1/\partial x_2 \partial x_3 -_n$$

$$-_n \partial v_3/\partial x_3 \times_n \partial v_3/\partial x_2 -_n v_3 \times_n \partial^2 v_3/\partial x_2 \partial x_3 +_n \partial v_3/\partial x_3 \times_n \partial v_2/\partial x_3 +_n$$

$$+_n v_3 \times_n \partial^2 v_2/\partial x_3^2)) =$$

$$= \partial f_3/\partial x_2 -_n \partial f_2/\partial x_3 +_n$$

$$+_n \alpha \times_n (\partial^3 v_3/\partial x_1^2 \partial x_2 +_n \partial^3 v_3/\partial x_2^3 +_n \partial^3 v_3/\partial x_2 \partial x_3^2 -_n \partial^3 v_2/\partial x_1^2 \partial x_3 -_n$$

$$-_n \partial^3 v_2/\partial x_2^2 \partial x_3 -_n \partial^3 v_2/\partial x_3^3)$$

   by coordinate x_2

$$\rho \times_n (\partial^2 v_1/\partial x_3 \partial t -_n \partial^2 v_3/\partial x_1 \partial t -_n$$

$$-_n(\partial v_1/\partial x_1 \times_n \partial v_3/\partial x_1 +_n v_1 \times_n \partial^2 v_3/\partial x_1^2 -_n \partial v_1/\partial x_1 \times_n \partial v_1/\partial x_3 -_n$$

$$-_n v_1 \times_n \partial^2 v_1/\partial x_1 \partial x_3 +_n$$

$$+_n \partial v_2/\partial x_1 \times_n \partial v_3/\partial x_2 +_n v_2 \times_n \partial^2 v_3/\partial x_1 \partial x_2 -_n \partial v_2/\partial x_1 \times_n \partial v_2/\partial x_3 -_n$$

$$-_n v_2 \times_n \partial^2 v_2/\partial x_1 \partial x_3 +_n$$

$$+_n \partial v_2/\partial x_3 \times_n \partial v_2/\partial x_1 +_n v_2 \times_n \partial^2 v_2/\partial x_1 \partial x_3 -_n \partial v_2/\partial x_3 \times_n \partial v_1/\partial x_2 -_n$$

$$-_n v_2 \times_n \partial^2 v_1/\partial x_2 \partial x_3 +_n$$

$$+_n \partial v_3/\partial x_3 \times_n \partial v_3/\partial x_1 +_n v_3 \times_n \partial^2 v_3/\partial x_1 \partial x_3 -_n \partial v_3/\partial x_3 \times_n \partial v_1/\partial x_3 -_n$$

$$-_n v_3 \times_n \partial^2 v_1/\partial x_3{}^2)) =$$

$$= \partial f_1/\partial x_3 -_n \partial f_3/\partial x_1 +_n$$

$$+_n \alpha \times_n (\partial^3 v_1/\partial x_1{}^2 \partial x_3 +_n \partial^3 v_1/\partial x_2{}^2 \partial x_3 +_n \partial^3 v_1/\partial x_3{}^3 -_n \partial^3 v_3/\partial x_1{}^3 -_n$$

$$-_n \partial^3 v_3/\partial x_1 \partial x_2{}^2 -_n \partial^3 v_3/\partial x_1 \partial x_3{}^2)$$

by coordinate x_3

$$\rho \times_n (\partial^2 v_2/\partial x_1 \partial t -_n \partial^2 v_1/\partial x_2 \partial t -_n$$

$$-_n(-\partial v_1/\partial x_1 \times_n \partial v_2/\partial x_1 -_n v_1 \times_n \partial^2 v_2/\partial x_1{}^2 +_n \partial v_1/\partial x_1 \times_n \partial v_1/\partial x_2 +_n$$

$$+_n v_1 \times_n \partial^2 v_1/\partial x_1 \partial x_2 +_n$$

$$+_n \partial v_3/\partial x_1 \times_n \partial v_3/\partial x_2 +_n v_3 \times_n \partial^2 v_3/\partial x_1 \partial x_2 -_n \partial v_3/\partial x_1 \times_n \partial v_2/\partial x_3 -_n$$

$$-_n v_3 \times_n \partial^2 v_2/\partial x_1 \partial x_3 -_n$$

$$-_n \partial v_2/\partial x_2 \times_n \partial v_2/\partial x_1 -_n v_2 \times_n \partial^2 v_2/\partial x_1 \partial x_2 +_n \partial v_2/\partial x_2 \times_n \partial v_1/\partial x_2 +_n$$

$$+_n v_2 \times_n \partial^2 v_1/\partial x_2{}^2 -_n$$

$$-_n \partial v_3/\partial x_2 \times_n \partial v_3/\partial x_1 -_n v_3 \times_n \partial^2 v_3/\partial x_1 \partial x_2 +_n \partial v_3/\partial x_2 \times_n \partial v_1/\partial x_3 +_n$$

$$+_n v_3 \times_n \partial^2 v_1/\partial x_2 \partial x_3)) =$$

$$= \partial f_2/\partial x_1 -_n \partial f_1/\partial x_2 +_n$$

$$+_n \alpha \times_n (\partial^3 v_2/\partial x_1{}^3 +_n \partial^3 v_2/\partial x_1 \partial x_2{}^2 +_n \partial^3 v_2/\partial x_1 \partial x_3{}^2 -_n \partial^3 v_1/\partial x_1{}^2 \partial x_2 -_n$$

$$-_n \partial^3 v_1/\partial x_2{}^3 -_n \partial^3 v_1/\partial x_2 \partial x_3{}^2)$$

is less than 1.

Let's consider particular case for initial conditions:

$$\mathbf{v}(\mathbf{x}, 0) = \mathbf{v}^0(\mathbf{x}) = (\mathbf{v}^0{}_1, \mathbf{v}^0{}_2, \mathbf{v}^0{}_3) = (x_1, -2 \times_n x_2, x_3)$$

We get

$$div\mathbf{v}^0 = (\nabla, \mathbf{v}^0) = \partial \mathbf{v}^0{}_1/\partial x_1 +_n \partial \mathbf{v}^0{}_2/\partial x_2 +_n \partial \mathbf{v}^0{}_3/\partial x_3 = 1 -_n 2 +_n 1 = 0$$

i.e. the initial conditions $\mathbf{v}^0(\mathbf{x})$ is a given divergence-free vector. For physically reasonable solutions, in classic Mathematics case we want to be sure $\mathbf{v}(\mathbf{x}, t)$ does not grow large as $|\mathbf{x}| \longrightarrow \infty$.

So, initial conditions $\mathbf{v}(\mathbf{x}, 0)$ in Observer's Mathematics and in our case are satisfied:

$$\partial \mathbf{v}^0{}_i/\partial x_j = \begin{cases} 1, \text{ if } i = 1, j = 1; i = 3, j = 3 \\ -2, \text{ if } i = 2, j = 2 \\ 0, \text{ for the rest pairs } (i, j) \end{cases}$$

$$|\partial^2 \mathbf{v}^0{}_i/\partial x_j \partial x_k| = 0$$

$$|\partial^3 \mathbf{v}^0{}_i/\partial x_j \partial x_k \partial x_l| = 0$$

where

$$i, j, k, l \in [1, 2, 3]$$

We are looking for solution of the system

$$\begin{cases} \rho \times_n (\partial/\partial t(\mathbf{rot}\ \mathbf{v}) -_n \mathbf{rot}\ (\mathbf{v} \times \mathbf{rot}\ \mathbf{v})) = \mathbf{rot}\ \mathbf{f} +_n \alpha \times_n \mathbf{rot}\Delta\mathbf{v} +_n \omega_{38} \\ div\mathbf{v} = \omega_{28} \\ \alpha > 0 \end{cases}$$

with given the initial conditions $\mathbf{v}^0(\mathbf{x})$.
Let's take

$$\mathbf{v}(\mathbf{x}, t) = (1 +_n t) \times_n \mathbf{v}^0(\mathbf{x}) = ((1 +_n t) \times_n \mathbf{v}^0{}_1, (1 +_n t) \times_n \mathbf{v}^0{}_2, (1 +_n t) \times_n \mathbf{v}^0{}_3) =$$

$$= ((1 +_n t) \times_n x_1, (1 +_n t) \times_n (-2 \times_n x_2), (1 +_n t) \times_n x_3)$$

We see that the initial conditions are satisfied.
Let's check second equation of the system:

$$div\mathbf{v} = (1 +_n t) \times_n div\mathbf{v}^0 +_n \varsigma_1 = \varsigma_1$$

where ς_1 is a random variable depends on n and m.
And we have in left-hand side and in right-hand side of the second equality two random variables having different distribution functions:

$$\varsigma_1 = \omega_{28}$$

Let's check first equation of the system, and start from coordinate x_1:

$$\rho \times_n (\partial^2((1 +_n t) \times_n x_3)/\partial x_2 \partial t -_n \partial^2((1 +_n t) \times_n (-2 \times_n x_2))/\partial x_3 \partial t$$
$$+_n \omega_{39}^1 +_n \omega_{39}^4 -_n$$

$$-_n(-\partial((1 +_n t) \times_n x_1)/\partial x_2 \times_n \partial((1 +_n t) \times_n x_3)/\partial x_1 -_n$$

$$-_n((1 +_n t) \times_n x_1) \times_n \partial^2((1 +_n t) \times_n x_3)/\partial x_1 \partial x_2 +_n$$

$$+_n \partial((1 +_n t) \times_n x_1)/\partial x_2 \times_n \partial((1 +_n t) \times_n x_1)/\partial x_3 +_n$$

$$+_n((1 +_n t) \times_n x_1) \times_n \partial^2((1 +_n t) \times_n x_1)/\partial x_2 \partial x_3 -_n$$

$$-_n \partial((1 +_n t) \times_n (-2 \times_n x_2))/\partial x_2 \times_n \partial((1 +_n t) \times_n x_3)/\partial x_2 -_n$$

$$-_n((1 +_n t) \times_n (-2 \times_n x_2)) \times_n \partial^2((1 +_n t) \times_n x_3)/\partial x_2^2 +_n$$

$$+_n \partial((1 +_n t) \times_n (-2 \times_n x_2))/\partial x_2 \times_n \partial((1 +_n t) \times_n (-2 \times_n x_2))/\partial x_3 +_n$$

$$+_n((1 +_n t) \times_n (-2 \times_n x_2)) \times_n \partial^2((1 +_n t) \times_n (-2 \times_n x_2))/\partial x_2 \partial x_3 +_n$$

$$+_n \partial((1 +_n t) \times_n x_1)/\partial x_3 \times_n \partial((1 +_n t) \times_n (-2 \times_n x_2))/\partial x_1 +_n$$

$$+_n((1 +_n t) \times_n x_1) \times_n \partial^2((1 +_n t) \times_n (-2 \times_n x_2))/\partial x_1 \partial x_3 -_n$$

$$-_n \partial((1 +_n t) \times_n x_1)/\partial x_3 \times_n \partial((1 +_n t) \times_n x_1)/\partial x_2 -_n$$

$$-_n((1 +_n t) \times_n x_1) \times_n \partial^2((1 +_n t) \times_n x_1)/\partial x_2 \partial x_3 -_n$$

$$-_n \partial((1 +_n t) \times_n x_3)/\partial x_3 \times_n \partial((1 +_n t) \times_n x_3)/\partial x_2 -_n$$

$$-_n((1 +_n t) \times_n x_3) \times_n \partial^2((1 +_n t) \times_n x_3)/\partial x_2 \partial x_3 +_n$$

$$+_n \partial((1 +_n t) \times_n x_3)/\partial x_3 \times_n \partial((1 +_n t) \times_n (-2 \times_n x_2))/\partial x_3 +_n$$

$$+_n((1 +_n t) \times_n x_3) \times_n \partial^2((1 +_n t) \times_n (-2 \times_n x_2))/\partial x_3^2 +_n$$

$$+_n \omega_{41}^1 +_n \omega_{42}^1 +_n \omega_{43}^1 +_n \omega_{44}^1)) =$$

$$= \partial f_3/\partial x_2 -_n \partial f_2/\partial x_3 +_n$$

$$+_n \alpha \times_n (\partial^3((1 +_n t) \times_n x_3)/\partial x_1^2 \partial x_2 +_n$$

$$+_n \partial^3((1 +_n t) \times_n x_3)/\partial x_2^3 +_n$$

$$+_n \partial^3((1 +_n t) \times_n x_3)/\partial x_2 \partial x_3^2 -_n$$

$$-_n \partial^3((1 +_n t) \times_n (-2 \times_n x_2))/\partial x_1^2 \partial x_3 -_n$$

$$-_n \partial^3((1 +_n t) \times_n (-2 \times_n x_2))/\partial x_2^2 \partial x_3 -_n$$

$$-_n \partial^3((1 +_n t) \times_n (-2 \times_n x_2))/\partial x_3^3 +_n \omega_{45}^1) +_n$$

$$+_n \omega_{38}^1$$

Let's assume now

$$\mathbf{f}(\mathbf{x}, t) = (f_1(x_1, x_2, x_3, t), f_2(x_1, x_2, x_3, t), f_3(x_1, x_2, x_3, t)) = \mathbf{const}$$

We have:

$$\rho \times_n (0 -_n 0 +_n \omega_{39}^1 +_n \omega_{39}^4 -_n$$

$$-_n (-0 -_n$$

$$-_n 0 +_n$$

$$+_n 0 +_n$$

$$+_n 0 -_n$$

$$-_n 0 -_n$$

$$-_n 0 +_n$$

$$+_n 0 +_n$$

$$+_n 0 +_n$$

$$+_n 0 +_n$$

$$+_n 0 -_n$$

$$-_n 0 -_n$$

$$-_n 0 -_n$$

$$-_n 0 -_n$$

$$-_n 0 +_n$$

$$+_n 0 +_n$$

$$+_n 0 +_n$$

$$+_n \omega_{41}^1 +_n \omega_{42}^1 +_n \omega_{43}^1 +_n \omega_{44}^1)) =$$

$$= 0 -_n 0 +_n$$

$$+_n \alpha \times_n (0 +_n$$

$$+_n 0 +_n$$

$$+_n 0 -_n$$

$$-_n 0 -_n$$

$$-_n 0 -_n$$

$$-_n 0 +_n \omega_{45}^1) +_n$$

$$+_n \omega_{38}^1$$

I.e.

$$\rho \times_n (\omega_{39}^1 +_n \omega_{39}^4 -_n (\omega_{41}^1 +_n \omega_{42}^1 +_n \omega_{43}^1 +_n \omega_{44}^1)) = \alpha \times_n \omega_{45}^1 +_n \omega_{38}^1$$

And we have in left-hand side and in right-hand side of this equality two random variables having different distribution functions.

Let's continue a check of first equation of the system, and go to coordinate x_2. Repeating process of calculations for coordinate x_1 we get

$$\rho \times_n (\omega_{39}^2 +_n \omega_{39}^5 +_n \omega_{41}^2 +_n \omega_{42}^2 +_n \omega_{43}^2 +_n \omega_{44}^2) =$$

$$= \alpha \times_n \omega_{45}^2 +_n \omega_{38}^2$$

And again we have in left-hand side and in right-hand side of this equality two random variables having different distribution functions. Let's go to coordinate x_3. Repeating process of calculations for coordinates x_1, x_2 we get

$$\rho \times_n (\omega_{39}^3 +_n \omega_{39}^6 +_n \omega_{41}^3 +_n \omega_{42}^3 +_n \omega_{43}^3 +_n \omega_{44}^3) =$$

$$= \alpha \times_n \omega_{45}^3 +_n \omega_{38}^3$$

And again we have in left-hand side and in right-hand side of this equality two random variables having different distribution functions.

So, the vector-function

$$\mathbf{v}(\mathbf{x}, t) = (1 +_n t) \times_n \mathbf{v}^0(\mathbf{x}) = ((1 +_n t) \times_n \mathbf{v}^0{}_1, (1 +_n t) \times_n \mathbf{v}^0{}_2, (1 +_n t) \times_n \mathbf{v}^0{}_3) =$$

$$= ((1 +_n t) \times_n x_1, (1 +_n t) \times_n (-2 \times_n x_2), (1 +_n t) \times_n x_3)$$

is the solution of the system

$$\begin{cases} \rho \times_n (\partial/\partial t(\mathrm{rot}\ \mathbf{v}) -_n \mathrm{rot}\ (\mathbf{v} \times \mathrm{rot}\ \mathbf{v})) = \mathrm{rot}\ \mathbf{f} +_n \alpha \times_n \mathrm{rot}\Delta\mathbf{v} +_n \omega_{38} \\ div\mathbf{v} = \omega_{28} \\ \alpha > 0 \end{cases}$$

with assumption

$$\mathbf{f}(\mathbf{x}, t) = (f_1(x_1, x_2, x_3, t), f_2(x_1, x_2, x_3, t), f_3(x_1, x_2, x_3, t)) = \mathbf{const}$$

and with accuracy up to distribution functions of some random variables depend on n and m, but not depend on any non-random solution, in particular,

$$\mathbf{v}(\mathbf{x}, t) = (v_1(x_1, x_2, x_3, t), v_2(x_1, x_2, x_3, t), v_3(x_1, x_2, x_3, t))$$

As physically reasonable solution, the solution of a system has to satisfy the bounded energy condition:

$$\int_e^g \int_c^d \int_a^b ((|(\mathbf{v}(\mathbf{x},t),\mathbf{v}(\mathbf{x},t))| \times_n \Delta x_1) \times_n \Delta x_2) \times_n \Delta x_3 \leq C$$

for any possible

$$a,b,c,d,e,g$$

with some

$$C = const$$

and for all

$$t \geq 0$$

But for point of view W_m – observer (just remind $m \geq n$) this condition is automatically satisfied because

$$\int_e^g \int_c^d \int_a^b ((|(\mathbf{v}(\mathbf{x},t),\mathbf{v}(\mathbf{x},t))| \times_n \Delta x_1) \times_n \Delta x_2) \times_n \Delta x_3 \in W_n$$

and so

$$\int_e^g \int_c^d \int_a^b ((|(\mathbf{v}(\mathbf{x},t),\mathbf{v}(\mathbf{x},t))| \times_n \Delta x_1) \times_n \Delta x_2) \times_n \Delta x_3 \leq \underbrace{9...9}_{n} \cdot \underbrace{9...9}_{n}$$

We can rewrite analogue of Navier-Stokes equations for incompressible fluid in coordinates

$$\mathbf{x} = (x_1, x_2, x_3), t \in W_n, t \geq 0$$

as follow:

$$\begin{cases} \rho \times_n (\partial v_i/\partial t +_n \sum_{j=1}^3 {}^n v_j \times_n \partial v_i/\partial x_j) = -\partial p/\partial x_i +_n f_i(\mathbf{x},t) +_n A_i \\ A_i = \alpha \times_n (\sum_{j=1}^3 {}^n \partial^2 v_i/\partial x_j^2) +_n \omega_{29}^i +_n \omega_{26}^i \\ div\mathbf{v} = \sum_{j=1}^3 {}^n \partial v_j/\partial x_j = \omega_{28} \\ i = 1,2,3 \\ \alpha > 0 \end{cases}$$

where we have unknown velocity vector

$$\mathbf{v}(\mathbf{x}, t) = (v_1(x_1, x_2, x_3, t), v_2(x_1, x_2, x_3, t), v_3(x_1, x_2, x_3, t)) \in E_3 W_n$$

and unknown function – pressure

$$p = p(\mathbf{x}, t) = p(x_1, x_2, x_3, t) \in W_n$$

We assume here that we have given externally applied force

$$\mathbf{f}(\mathbf{x}, t) = (f_1(x_1, x_2, x_3, t), f_2(x_1, x_2, x_3, t), f_3(x_1, x_2, x_3, t))$$
$$= \mathbf{const} = \mathbf{0} \in E_3 W_n$$

And we assume that all elements of this system belong to W_n. Let's consider same particular case for initial conditions:

$$\mathbf{v}(\mathbf{x}, 0) = \mathbf{v}^0(\mathbf{x}) = (\mathbf{v}^0{}_1, \mathbf{v}^0{}_2, \mathbf{v}^0{}_3) = (x_1, -2 \times_n x_2, x_3)$$

We get

$$div \mathbf{v}^0 = (\nabla, \mathbf{v}^0) = \partial \mathbf{v}^0{}_1/\partial x_1 +_n \partial \mathbf{v}^0{}_2/\partial x_2 +_n \partial \mathbf{v}^0{}_3/\partial x_3 = 1 -_n 2 +_n 1 = 0$$

i.e. the initial conditions $\mathbf{v}^0(\mathbf{x})$ is a given divergence-free vector. For physically reasonable solutions, in classical Mathematics case we want to be sure $\mathbf{v}(\mathbf{x}, t)$ does not grow large as $|\mathbf{x}| \longrightarrow \infty$. So, initial conditions $\mathbf{v}(\mathbf{x}, 0)$ in Mathematics with Observers and in our case are satisfied:

$$\partial \mathbf{v}^0{}_i/\partial x_j = \begin{cases} 1, \text{ if } i = 1, j = 1; i = 3, j = 3 \\ -2, \text{ if } i = 2, j = 2 \\ 0, \text{ for the rest pairs } (i, j) \end{cases}$$

$$|\partial^2 \mathbf{v}^0{}_i/\partial x_j \partial x_k| = 0$$

$$|\partial^3 \mathbf{v}^0{}_i/\partial x_j \partial x_k \partial x_l| = 0$$

where

$$i, j, k, l \in [1, 2, 3]$$

Let's take

$$\mathbf{v}(\mathbf{x}, t) = (1 +_n t) \times_n \mathbf{v}^0(\mathbf{x}) = ((1 +_n t) \times_n \mathbf{v}^0{}_1, (1 +_n t) \times_n \mathbf{v}^0{}_2, (1 +_n t) \times_n \mathbf{v}^0{}_3) =$$

$$= ((1 +_n t) \times_n x_1, (1 +_n t) \times_n (-2 \times_n x_2), (1 +_n t) \times_n x_3)$$

The unknown function - pressure

$$p = p(\mathbf{x}, t) = p(x_1, x_2, x_3, t) \in W_n$$

we'll find later. We see that the initial conditions are satisfied. Let's check second equation of the system:

$$div\mathbf{v} = (1 +_n t) \times_n div\mathbf{v}^0 +_n \varsigma_1 = \varsigma_1$$

And we have in left-hand side and in right-hand side of the second equality two random variables having different distribution functions:

$$\varsigma_1 = \omega_{28}$$

Let's check first equation of the system, and start from coordinate x_1:

$$\rho \times_n (\partial v_1/\partial t +_n \sum_{j=1}^{3} {}^n v_j \times_n \partial v_1/\partial x_j) = -\partial p/\partial x_1 +_n \alpha \times_n (\sum_{j=1}^{3} {}^n \partial^2 v_1/\partial x_j^2)$$
$$+_n \omega_{29}^1 +_n \omega_{26}^1$$

We have

$$v_1 = (1 +_n t) \times_n x_1$$

$$\partial v_1/\partial t = \partial((1 +_n t) \times_n x_1)/\partial t = x_1 +_n \varsigma_2$$

$$\partial v_1/\partial x_1 = \partial((1 +_n t) \times_n x_1)/\partial x_1 = 1 +_n t +_n \varsigma_3$$

$$\partial v_1/\partial x_2 = \partial((1 +_n t) \times_n x_1)/\partial x_2 = 0$$

$$\partial v_1/\partial x_3 = \partial((1 +_n t) \times_n x_1)/\partial x_3 = 0$$

$$\partial^2 v_1/\partial x_1^2 = \partial^2((1 +_n t) \times_n x_1)/\partial x_1^2 = \varsigma_4$$

$$\partial^2 v_1/\partial x_2^2 = \partial^2((1 +_n t) \times_n x_1)/\partial x_2^2 = 0$$

$$\partial^2 v_1/\partial x_3^2 = \partial^2((1 +_n t) \times_n x_1)/\partial x_3^2 = 0$$

where $\varsigma_2, \varsigma_3, \varsigma_4$ are the random variables depend on n and m. So, we can rewrite first equation of the system for coordinate x_1:

$$\rho \times_n (x_1 +_n \varsigma_2 +_n ((1 +_n t) \times_n x_1) \times_n (1 +_n t +_n \varsigma_3)) = -\partial p/\partial x_1 +_n \varsigma_4 +_n \omega_{29}^1 +_n \omega_{26}^1$$

Let's now check first equation of the system with coordinate x_2:

$$\rho \times_n (\partial v_2/\partial t +_n \sum_{j=1}^{3} {}^n v_j \times_n \partial v_2/\partial x_j) = -\partial p/\partial x_2 +_n \alpha \times_n (\sum_{j=1}^{3} {}^n \partial^2 v_2/\partial x_j^2)$$
$$+_n \omega_{29}^2 +_n \omega_{26}^2$$

We have

$$v_2 = (1 +_n t) \times_n (-2 \times_n x_2)$$

$$\partial v_2/\partial t = \partial(((1 +_n t) \times_n (-2 \times_n x_2))/\partial t = -2 \times_n x_2 +_n \varsigma_5$$

$$\partial v_2/\partial x_1 = \partial((1 +_n t) \times_n (-2 \times_n x_2))/\partial x_1 = 0$$

$$\partial v_2/\partial x_2 = \partial((1 +_n t) \times_n (-2 \times_n x_2))/\partial x_2 = -2 \times_n (1 +_n t) +_n \varsigma_6$$

$$\partial v_2/\partial x_3 = \partial((1 +_n t) \times_n (-2 \times_n x_2))/\partial x_3 = 0$$

$$\partial^2 v_2/\partial x_1^2 = \partial^2((1 +_n t) \times_n (-2 \times_n x_2))/\partial x_1^2 = 0$$

$$\partial^2 v_2/\partial x_2^2 = \partial^2((1 +_n t) \times_n (-2 \times_n x_2))/\partial x_2^2 = \varsigma_7$$

$$\partial^2 v_2/\partial x_3^2 = \partial^2((1 +_n t) \times_n (-2 \times_n x_2))/\partial x_3^2 = 0$$

where $\varsigma_5, \varsigma_6, \varsigma_7$ are the random variables depend on n and m.
So, we can rewrite first equation of the system for coordinate x_2:

$$\rho \times_n (-2 \times_n x_2 +_n \varsigma_5 +_n ((1 +_n t) \times_n (-2 \times_n x_2)) \times_n (-2 \times_n (1 +_n t) +_n \varsigma_6)) =$$

$$= -\partial p/\partial x_2 +_n \varsigma_7 +_n \omega_{29}^2 +_n \omega_{26}^2$$

Let's now check first equation of the system with coordinate x_3:

$$\rho \times_n (\partial v_3/\partial t +_n \sum_{j=1}^{3} {}^n v_j \times_n \partial v_3/\partial x_j) = -\partial p/\partial x_3 +_n \alpha \times_n (\sum_{j=1}^{3} {}^n \partial^2 v_3/\partial x_j^2)$$
$$+_n \omega_{29}^3 +_n \omega_{26}^3$$

We have

$$v_3 = (1 +_n t) \times_n x_3$$

$$\partial v_3 / \partial t = \partial((1 +_n t) \times_n x_3) / \partial t = x_3 +_n \varsigma_8$$

$$\partial v_3 / \partial x_1 = \partial((1 +_n t) \times_n x_3) / \partial x_1 = 0$$

$$\partial v_3 / \partial x_2 = ((1 +_n t) \times_n x_3) / \partial x_2 = 0$$

$$\partial v_3 / \partial x_3 = \partial((1 +_n t) \times_n x_3) / \partial x_3 = 1 +_n t + \varsigma_9$$

$$\partial^2 v_3 / \partial x_1^2 = \partial^2((1 +_n t) \times_n x_3) / \partial x_1^2 = 0$$

$$\partial^2 v_3 / \partial x_2^2 = \partial^2((1 +_n t) \times_n x_3) / \partial x_2^2 = 0$$

$$\partial^2 v_3 / \partial x_3^2 = \partial^2((1 +_n t) \times_n x_3) / \partial x_3^2 = \varsigma_{10}$$

where $\varsigma_8, \varsigma_9, \varsigma_{10}$ are the random variables depend on n and m.
So, we can rewrite first equation of the system for coordinate x_3:

$$\rho \times_n (x_3 +_n \varsigma_8 +_n ((1 +_n t) \times_n x_3) \times_n ((1 +_n t) +_n \varsigma_9)) = -\partial p / \partial x_3 +_n \varsigma_{10} +_n \omega_{29}^3 +_n \omega_{26}^3$$

And now we get the system of equations for the pressure $p = p(x_1, x_2, x_3, t)$ definition:

$$\begin{cases} \partial p / \partial x_1 = -\rho \times_n (x_1 +_n \varsigma_2 +_n ((1 +_n t) \times_n x_1) \times_n (1 +_n t +_n \varsigma_3)) +_n E \\ E = \varsigma_{11} +_n \omega_{29}^1 +_n \omega_{26}^1 \\ G = -2 \times_n x_2 +_n \varsigma_5 +_n ((1 +_n t) \times_n (2 \times_n x_2)) \times_n (2 \times_n (1 +_n t) -_n \varsigma_6) \\ \partial p / \partial x_2 = -\rho \times_n G +_n \varsigma_{12} +_n \omega_{29}^2 +_n \omega_{26}^2 \\ \partial p / \partial x_3 = -\rho \times_n (x_3 +_n \varsigma_8 +_n ((1 +_n t) \times_n x_3) \times_n ((1 +_n t) +_n \varsigma_9)) +_n H \\ H = \varsigma_{13} +_n \omega_{29}^3 +_n \omega_{26}^3 \end{cases}$$

where $\varsigma_{11}, \varsigma_{12}, \varsigma_{13}$ are the random variables depend on n and m, and

$$\varsigma_{11} = \alpha \times_n \varsigma_4, \varsigma_{12} = \alpha \times_n \varsigma_7, \varsigma_{13} = \alpha \times_n \varsigma_{10}$$

Many times applying random variables δ_2, δ_3 we can rewrite this system as the following:

$$\begin{cases} \partial p / \partial x_1 = -\rho \times_n ((1 +_n t +_n (1 +_n t)^2) \times_n x_1) +_n \varsigma_{14} +_n \omega_{29}^1 +_n \omega_{26}^1 \\ \partial p / \partial x_2 = -\rho \times_n ((-4 \times_n (1 +_n t) +_n 4 \times_n (1 +_n t)^2) \times_n x_2) +_n \varsigma_{15} +_n \omega_{29}^2 +_n \omega_{26}^2 \\ \partial p / \partial x_3 = -\rho \times_n ((1 +_n t +_n (1 +_n t)^2) \times_n x_3) +_n \varsigma_{16} +_n \omega_{29}^3 +_n \omega_{26}^3 \end{cases}$$

where $\varsigma_{14}, \varsigma_{15}, \varsigma_{16}$ are the random variables depend on n and m.
Now we get

$$p = p(x_1, x_2, x_3, t) = -\rho \times_n ((1 +_n t +_n (1 +_n t)^2) \times_n (\frac{1}{2} \times_n (x_1 \times_n x_1 -_n x_1^0 \times_n x_1^0) -_n$$

$$-_n \Delta x_1 \times_n (\frac{1}{2} \times_n (x_1 -_n x_1^0))) +_n \varsigma_{17} +_n$$

$$+_n ((-4 \times_n (1 +_n t) +_n 4 \times_n (1 +_n t)^2) \times_n (\frac{1}{2} \times_n (x_2 \times_n x_2 -_n x_2^0 \times_n x_2^0) -_n$$

$$-_n \Delta x_2 \times_n (\frac{1}{2} \times_n (x_2 -_n x_2^0))) +_n \varsigma_{18} +_n$$

$$+_n ((1 +_n t +_n (1 +_n t)^2) \times_n (\frac{1}{2} \times_n (x_3 \times_n x_3 -_n x_3^0 \times_n x_3^0) -_n$$

$$-_n \Delta x_3 \times_n (\frac{1}{2} \times_n (x_3 -_n x_3^0))) +_n \varsigma_{19} +_n$$

$$+_n p(x_1^0, x_2^0, x_3^0, t)$$

The solution of a system has to satisfy the bounded energy condition:

$$\int_e^g \int_c^d \int_a^b ((|(\mathbf{v}(\mathbf{x}, t), \mathbf{v}(\mathbf{x}, t))| \times_n \Delta x_1) \times_n \Delta x_2) \times_n \Delta x_3 \leq C$$

for any possible

$$a, b, c, d, e, g$$

˜ with some

$$C = const$$

and for all

$$t \geq 0$$

But for point of view W_m-observer ($m > n$) this condition is automatically satisfied because

$$\int_e^g \int_c^d \int_a^b ((|(\mathbf{v}(\mathbf{x}, t), \mathbf{v}(\mathbf{x}, t))| \times_n \Delta x_1) \times_n \Delta x_2) \times_n \Delta x_3 \in W_n$$

and so

$$\int_e^g \int_c^d \int_a^b ((|(\mathbf{v}(\mathbf{x},t), \mathbf{v}(\mathbf{x},t))| \times_n \Delta x_1) \times_n \Delta x_2) \times_n \Delta x_3 \le \underbrace{9...9}_{n} \cdot \underbrace{9...9}_{n}$$

Also the solution $p = p(x_1, x_2, x_3, t)$ of a system satisfies the pressure bounded condition:

$$\int_e^g \int_c^d \int_a^b ((|p(x_1, x_2, x_3, t)| \times_n \Delta x_1) \times_n \Delta x_2) \times_n \Delta x_3 \le J$$

for any possible

$$a, b, c, d, e, g$$

with some

$$J = const$$

and for all

$$t \ge 0$$

But for point of view W_m – observer ($m > n$) this condition is automatically satisfied because

$$\int_e^g \int_c^d \int_a^b ((|p(x_1, x_2, x_3, t)| \times_n \Delta x_1) \times_n \Delta x_2) \times_n \Delta x_3 \in W_n$$

and so

$$\int_e^g \int_c^d \int_a^b ((|p(x_1, x_2, x_3, t)| \times_n \Delta x_1) \times_n \Delta x_2) \times_n \Delta x_3 \le \underbrace{9...9}_{n} \cdot \underbrace{9...9}_{n}$$

We showed examples of breakdown of Navier-Stokes solutions for 3-dimensional case in Mathematics with Observers, i.e. in situation

$$\mathbf{f}(\mathbf{x}, t) = (f_1(x_1, x_2, x_3, t), f_2(x_1, x_2, x_3, t), f_3(x_1, x_2, x_3, t))$$
$$= \mathbf{const} = \mathbf{0} \in E_3 W_n$$

the solutions of Navier-Stokes equations

$$\mathbf{v} = \mathbf{v}(\mathbf{x}, t)$$

and

$$p = p(\mathbf{x}, t)$$

with divergence-free initial conditions

$$\mathbf{v}(\mathbf{x}, 0) = \mathbf{v}^0(\mathbf{x}) = (\mathbf{v}^0{}_1, \mathbf{v}^0{}_2, \mathbf{v}^0{}_3)$$

cannot be determined – not random – functions. The same statement takes a place not for particular cases only, but in general cases:

Theorem 11.31 *For any not random divergence-free initial conditions, for any not random externally applied force Navier-Stokes equations in Mathematics with Observers do not have any not random solutions, i.e. we get solutions breakdown of Navier-Stokes equations in Mathematics with Observers for 3-dimensional case. We mean here both expressions of these equations for incompressible fluid:*

$$\begin{cases} \rho \times_n \left(\partial v_i / \partial t +_n \sum_{j=1}^{3} {}^n v_j \times_n \partial v_i / \partial x_j \right) = -\partial p / \partial x_i +_n f_i(\boldsymbol{x}, t) +_n A_i \\ A_i = \alpha \times_n \left(\sum_{j=1}^{3} {}^n \partial^2 v_i / \partial x_j^2 \right) +_n \omega_{29}^i +_n \omega_{26}^i \\ div\boldsymbol{v} = \sum_{j=1}^{3} {}^n \partial v_j / \partial x_j = \omega_{28} \\ i = 1, 2, 3 \\ \alpha > 0 \end{cases}$$

and

$$\begin{cases} \rho \times_n \left(\partial / \partial t(\boldsymbol{rot}\ \boldsymbol{v}) -_n \boldsymbol{rot}\ (\boldsymbol{v} \times \boldsymbol{rot}\ \boldsymbol{v}) \right) = \boldsymbol{rot}\ \boldsymbol{f} +_n \alpha \times_n \boldsymbol{rot} \Delta \boldsymbol{v} +_n \omega_{38} \\ div\boldsymbol{v} = \omega_{28} \\ \alpha > 0 \end{cases}$$

Proof:
Random variables and vectors appearing when we derive Euler and Navier-Stokes equations in Mathematics with Observers

$$\omega_1, \omega_2, \ldots, \omega_{45}$$

and random variables and vectors appearing when we calculate derivatives and integrals

$$\xi_1, \xi_2, \ldots, \xi_{73}; \iota_1, \iota_2, \ldots, \iota_{34}$$

have different distribution functions and can not be mutually cut.

12

Observability and Relativistic Fluid Mechanics

The governing principles in Fluid Mechanics are the conservation laws for mass, momentum, and energy. And in classic Physics and Mathematics the conservation laws characterizing special relativistic fluid mechanics are invariant (in fact co-variant) under Poincare group of transformations (see [7], [6], [11]).

Let's consider this situation from Mathematics with Observers point of view. First of all let's consider general question – Operations over matrices in Mathematics with Observers.

Let's

$$A = (a_{ij}), B = (b_{ij}), i \in (1, 2, \ldots, k), j \in (1, 2, \ldots, m)$$

We put

$$A +_n B = (a_{ij}) +_n (b_{ij}) = (a_{ij} +_n b_{ij})$$

We assume that all elements of this equality belong to W_n.
If

$$A = (a_{ij}), B = (b_{pq}),$$

$$i \in (1, 2, \ldots, k), j \in (1, 2, \ldots, m), p \in (1, 2, \ldots, r), q \in (1, 2, \ldots, s)$$

and $m = r$, then we define

$$A \times_n B = D = (d_{iq} = (\Sigma_{j=1}^{m} a_{ij} \times_n b_{jq})$$

We assume that all elements of this equality and summation belong to W_n.
If α is a scalar, we define

$$\alpha \times_n A = (\alpha \times_n a_{ij})$$

So, we get as it is above (for example, for matrices A, B, D sized $k \times k$):
1. The probability of equality

$$A \times_n (B +_n D) = A \times_n B +_n A \times_n D$$

DOI: 10.1201/9781003175902-12

is less than 1.

2. The probability of equality

$$A \times_n (B \times_n D) = (A \times_n B) \times_n D$$

is less than 1.

3. The probability of equality

$$\alpha \times_n (B +_n D) = \alpha \times_n B +_n \alpha \times_n D$$

is less than 1.

4. The probability of equality

$$\alpha \times_n (\beta \times_n D) = (\alpha \times_n \beta) \times_n D$$

is less than 1 (β is a scalar).

First of all we'll go to General, Orthogonal, Lorentz and Poincare matrix groups considered in Mathematics with Observers. General matrix group Lie **GL** (p, W_n) is the set of all invertible $p \times p$ matrices with entries $\in W_n$. For our applications we will consider below only cases $p = 3$, $p = 4$ or $p = 5$. So, we continue our analysis for these p.

GL $(3, W_n)$ has identity:

$$1 = \begin{bmatrix} 1 & 0 & 0 \\ 0 & 1 & 0 \\ 0 & 0 & 1 \end{bmatrix}$$

And we have

$$1^{-1} = 1$$

Also

$$A \times_n 1 = 1 \times_n A = A$$

for any 3×3 matrix A.

Let's consider now $n = 2$ and consider three matrices $A, B, C \in$ **GL** $(3, W_n)$:

$$A = \begin{bmatrix} 0.2 & 0 & 0 \\ 0 & 1 & 0 \\ 0 & 0 & 1 \end{bmatrix}$$

$$B = \begin{bmatrix} 0.25 & 0 & 0 \\ 0 & 1 & 0 \\ 0 & 0 & 1 \end{bmatrix}$$

$$C = \begin{bmatrix} 0.5 & 0 & 0 \\ 0 & 1 & 0 \\ 0 & 0 & 1 \end{bmatrix}$$

These matrices are invertible, and each of them has ten inverses:

$$A^{-1} = \begin{bmatrix} 5.00 & 0 & 0 \\ 0 & 1 & 0 \\ 0 & 0 & 1 \end{bmatrix}$$

....

$$A^{-1} = \begin{bmatrix} 5.09 & 0 & 0 \\ 0 & 1 & 0 \\ 0 & 0 & 1 \end{bmatrix}$$

$$B^{-1} = \begin{bmatrix} 4.00 & 0 & 0 \\ 0 & 1 & 0 \\ 0 & 0 & 1 \end{bmatrix}$$

....

$$B^{-1} = \begin{bmatrix} 4.09 & 0 & 0 \\ 0 & 1 & 0 \\ 0 & 0 & 1 \end{bmatrix}$$

$$C^{-1} = \begin{bmatrix} 2.00 & 0 & 0 \\ 0 & 1 & 0 \\ 0 & 0 & 1 \end{bmatrix}$$

....

$$C^{-1} = \begin{bmatrix} 2.09 & 0 & 0 \\ 0 & 1 & 0 \\ 0 & 0 & 1 \end{bmatrix}$$

We get

$$A \times_n B = \begin{bmatrix} 0.04 & 0 & 0 \\ 0 & 1 & 0 \\ 0 & 0 & 1 \end{bmatrix}$$

$$(A \times_n B) \times_n C = \begin{bmatrix} 0.00 & 0 & 0 \\ 0 & 1 & 0 \\ 0 & 0 & 1 \end{bmatrix}$$

and

$$B \times_n C = \begin{bmatrix} 0.10 & 0 & 0 \\ 0 & 1 & 0 \\ 0 & 0 & 1 \end{bmatrix}$$

$$A \times_n (B \times_n C) = \begin{bmatrix} 0.02 & 0 & 0 \\ 0 & 1 & 0 \\ 0 & 0 & 1 \end{bmatrix}$$

That means

$$(A \times_n B) \times_n C \neq A \times_n (B \times_n C)$$

i.e. multiplicative associativity may fail. So, $\mathbf{GL}\ (p, W_n)$ is not a group in Mathematics with Observers for $p = 3$. We can state the same and for $p = 4$ and for $p = 5$ and for any n.

From another side let's consider again $n = 2$ and consider three another matrices

$$A', B', C' \in \mathbf{GL}\ (3, W_n) : A' = \begin{bmatrix} 2 & 0 & 0 \\ 0 & 2 & 0 \\ 0 & 0 & -2 \end{bmatrix}$$

$$B' = \begin{bmatrix} 2 & 0 & 0 \\ 0 & -2 & 0 \\ 0 & 0 & 2 \end{bmatrix}$$

$$C' = \begin{bmatrix} -2 & 0 & 0 \\ 0 & 2 & 0 \\ 0 & 0 & 2 \end{bmatrix}$$

These matrices are invertible, and each of them has exactly one inverse:

$$A'^{-1} = \begin{bmatrix} 0.50 & 0 & 0 \\ 0 & 0.50 & 0 \\ 0 & 0 & -0.50 \end{bmatrix}$$

$$B'^{-1} = \begin{bmatrix} 0.50 & 0 & 0 \\ 0 & -0.50 & 0 \\ 0 & 0 & 0.50 \end{bmatrix}$$

$$C'^{-1} = \begin{bmatrix} -0.50 & 0 & 0 \\ 0 & 0.50 & 0 \\ 0 & 0 & 0.50 \end{bmatrix}$$

We get

$$A' \times_n B' = \begin{bmatrix} 4.00 & 0 & 0 \\ 0 & -4.00 & 0 \\ 0 & 0 & -4.00 \end{bmatrix}$$

$$(A' \times_n B') \times_n C' = \begin{bmatrix} -8.00 & 0 & 0 \\ 0 & -8.00 & 0 \\ 0 & 0 & -8.00 \end{bmatrix}$$

and

$$B' \times_n C' = \begin{bmatrix} -4.00 & 0 & 0 \\ 0 & -4.00 & 0 \\ 0 & 0 & 4.00 \end{bmatrix}$$

$$A' \times_n (B' \times_n C') = \begin{bmatrix} -8.00 & 0 & 0 \\ 0 & -8.00 & 0 \\ 0 & 0 & -8.00 \end{bmatrix}$$

That means

$$(A \times_n B) \times_n C = A \times_n (B \times_n C)$$

i.e. multiplicative associativity takes a place in this case.
So, the probabilities of equality

$$(A \times_n B) \times_n C = A \times_n (B \times_n C)$$

and uniqueness of inverse matrices is less than 1 for the set **GL** $(3, W_n)$. We can state the same and for $p = 4$ and for $p = 5$ and for any possible n.
Let's continue to consider $n = 2$ and take matrix $D \in \mathbf{GL}\ (3, W_n)$:

$$D = \begin{bmatrix} 3 & 0 & 0 \\ 0 & 1 & 0 \\ 0 & 0 & 1 \end{bmatrix}$$

In this case matrix D^{-1} does not exist. Let's note in classic case $det D \neq 0$, and D^{-1} has to exist.
Let's consider matrix $E \in \mathbf{GL}\ (3, W_n)$:

$$E = \begin{bmatrix} 0.25 & 0 & 0 \\ 0 & 0.25 & 0 \\ 0 & 0 & 0.25 \end{bmatrix}$$

We get

$$E^{-1} = \begin{bmatrix} 4.00 & 0 & 0 \\ 0 & 4.00 & 0 \\ 0 & 0 & 4.00 \end{bmatrix}$$

So, $det E = 0$, and E is invertible matrix. Moreover, from point of view W_m- observer with $(m \geq 4)$ matrix E has $1,000$ different inverse matrices, receiving by substitutions of numbers 4 on diagonal to any number from set $[4.00, 4.01, \ldots, 4.09]$:

$$E^{-1} = \begin{bmatrix} 4.01 & 0 & 0 \\ 0 & 4.00 & 0 \\ 0 & 0 & 4.00 \end{bmatrix}$$

·······················

$$E^{-1} = \begin{bmatrix} 4.09 & 0 & 0 \\ 0 & 4.09 & 0 \\ 0 & 0 & 4.09 \end{bmatrix}$$

And again let's note in classic case $det E = 0$ means E^{-1} does not exist.
Let's consider now general determinant question for **GL** $(3, W_n)$, **GL** $(4, W_n)$ and **GL** $(5, W_n)$: First of all note we mean "determinant of matrix A" as a classic combination of sums and products of elements $\in A$ calculated by pairs from left to right.
For $p = 3$ and $A = (a_{ij})$; $i, j = 1, 2, 3$

$$det A = ((((((a_{11} \times_n a_{22}) \times_n a_{33} +_n (a_{21} \times_n a_{32}) \times_n a_{13}) +_n (a_{12} \times_n a_{23}) \times_n a_{31}) -_n$$

$$-_n (a_{13} \times_n a_{22}) \times_n a_{13}) -_n (a_{21} \times_n a_{12}) \times_n a_{33}) -_n (a_{32} \times_n a_{23}) \times_n a_{11})$$

Analogous pairing is used for $p = 4$ and $p = 5$. Note that there are other ways for pairing, and generally for different pairings results of determinant calculations will be different, and difference between any two of them is a random variable depends on n and m.

For $A, B \in \textbf{GL}\ (3, W_n)$ is equality

$$det A \times_n det B = det(A \times_n B)$$

correct or not? Same question - for $\textbf{GL}\ (4, W_n)$ and for $\textbf{GL}\ (5, W_n)$. Let's again take $n = 2$ and matrices E and E^{-1}:

$$E = \begin{bmatrix} 0.25 & 0 & 0 \\ 0 & 0.25 & 0 \\ 0 & 0 & 0.25 \end{bmatrix}$$

and

$$E^{-1} = \begin{bmatrix} 4.00 & 0 & 0 \\ 0 & 4.00 & 0 \\ 0 & 0 & 4.00 \end{bmatrix}$$

We have

$$E \times_n E^{-1} = \textbf{1}$$

and

$$det E = 0;\ det E^{-1} = 64;\ det\textbf{1} = 1;\ det E \times_n det E^{-1} \neq det\textbf{1}$$

That means answer for question in this case is negative.

From another side let's consider again $n = 2$ and consider again two matrices

$A', B' \in \textbf{GL}\ (3, W_n):$

$$A' = \begin{bmatrix} 2 & 0 & 0 \\ 0 & 2 & 0 \\ 0 & 0 & -2 \end{bmatrix}$$

$$B' = \begin{bmatrix} 2 & 0 & 0 \\ 0 & -2 & 0 \\ 0 & 0 & 2 \end{bmatrix}$$

These matrices are invertible, and each of them has exactly one inverse.

We get

$$A' \times_n B' = \begin{bmatrix} 4.00 & 0 & 0 \\ 0 & -4.00 & 0 \\ 0 & 0 & -4.00 \end{bmatrix}$$

and

$$det\, A' \times_n det\, B' = det(A' \times_n B') = 64$$

That means answer for question in this case is positive. So, the probability of equality

$$det\, A \times_n det\, B = det(A \times_n B)$$

for any matrices $\in$ **GL** $(3, W_n)$ is less than 1. Same answer takes a place for any n and for $p = 4$ and for $p = 5$.

Generally speaking, the sets **GL** $(3, W_n)$, **GL** $(4, W_n)$ and **GL** $(5, W_n)$ are not the groups in Mathematics with Observers, and group definition's conditions take a place here with some probability less than 1.

So, we proved

Theorem 12.1 *The sets* **GL** $(3, W_n)$, **GL** $(4, W_n)$ *and* **GL** $(5, W_n)$ *are not the groups in Mathematics with Observers, and group definition's conditions take a place here with some probability less than 1.*

In Mathematics with Observers we call matrix $A \in$ **GL** $(3, W_n)$ is orthogonal matrix if it's columns and rows are orthogonal unit vectors, i.e.

$$A^T \times_n A = A \times_n A^T = 1$$

We call the set of these matrices **O** (3) the orthogonal Lie group. And we have

$$1^{-1} = 1^T = 1$$

where upper index T means transpose. Also

$$A \times_n 1 = 1 \times_n A = A$$

for any $A \in$ **O** (3). That means a matrix A is orthogonal if its transpose is its inverse:

$$A^T = A^{-1}$$

But orthogonal matrix may have more than one inverse.

Let's take $n = 2$ and matrix A:

$$A = \begin{bmatrix} 0.81 & 0.69 & 0 \\ -0.69 & 0.81 & 0 \\ 0 & 0 & 1 \end{bmatrix}$$

We have

$$A^{-1} = \begin{bmatrix} 0.80 & -0.60 & 0 \\ 0.60 & 0.80 & 0 \\ 0 & 0 & 1 \end{bmatrix}$$

$$A^{-1} = \begin{bmatrix} 0.80 & -0.61 & 0 \\ 0.61 & 0.80 & 0 \\ 0 & 0 & 1 \end{bmatrix}$$

....

$$A^{-1} = \begin{bmatrix} 0.80 & -0.69 & 0 \\ 0.69 & 0.80 & 0 \\ 0 & 0 & 1 \end{bmatrix}$$

$$A^{-1} = \begin{bmatrix} 0.81 & -0.60 & 0 \\ 0.60 & 0.81 & 0 \\ 0 & 0 & 1 \end{bmatrix}$$

....

$$A^{-1} = \begin{bmatrix} 0.81 & -0.69 & 0 \\ 0.69 & 0.81 & 0 \\ 0 & 0 & 1 \end{bmatrix}$$

....

$$A^{-1} = \begin{bmatrix} 0.89 & -0.69 & 0 \\ 0.69 & 0.89 & 0 \\ 0 & 0 & 1 \end{bmatrix}$$

Total matrix A has 100 inverses from point of view W_m- observer with $m \geq 3$. Let's take two column vectors $\mathbf{v}, \mathbf{w} \in E_3 W_n$ and orthogonal matrix $A \in \mathbf{O}\,(3)$.

We consider case $n = 2$, and first let's

$$A = \begin{bmatrix} 0.81 & 0.69 & 0 \\ -0.69 & 0.81 & 0 \\ 0 & 0 & 1 \end{bmatrix}$$

We get

$$A \times_n \mathbf{v} = \begin{bmatrix} 0.81 \times_2 v^1 +_n 0.69 \times_2 v^2 \\ -0.69 \times_2 v^1 +_n 0.81 \times_2 v^2 \\ 1 \times_2 v^3 \end{bmatrix}$$

$$A \times_n \mathbf{w} = \begin{bmatrix} 0.81 \times_2 w^1 +_n 0.69 \times_2 w^2 \\ -0.69 \times_2 w^1 +_n 0.81 \times_2 w^2 \\ 1 \times_2 w^3 \end{bmatrix}$$

And

$$(A \times_n \mathbf{v}, A \times_n \mathbf{w}) = (A \times_n \mathbf{v})^T \times_n (A \times_n \mathbf{w}) =$$

$$= (0.81 \times_2 v^1 +_n 0.69 \times_2 v^2) \times_2 (0.81 \times_2 w^1 +_n 0.69 \times_2 w^2) +_2$$

$$+_2 (-0.69 \times_2 v^1 +_n 0.81 \times_2 v^2) \times_2 (-0.69 \times_2 w^1 +_n 0.81 \times_2 w^2) +_2 v^3 \times_2 w^3$$

If we take

$$\mathbf{v}^T = (1.14, -2.81, 3.47); \mathbf{w}^T = (-2.64, 0.73, 1.29)$$

we get

$$((A \times_n \mathbf{v})^T, (A \times_n \mathbf{w})^T) = ((-0.97, -3.01, 3.47), (-1.68, 2.30, 1.29)) =$$

$$= 1.51 -_2 6.92 +_2 4.42 = -0.99$$

and

$$(\mathbf{v}^T, \mathbf{w}^T) = -2.98 -_2 2.02 +_2 4.42 = -0.58$$

i.e.

$$((A \times_n \mathbf{v})^T, (A \times_n \mathbf{w})^T) \neq (\mathbf{v}^T, \mathbf{w}^T)$$

We continue to consider case $n = 2$, and now let's

$$A = \begin{bmatrix} 0 & 1 & 0 \\ 1 & 0 & 0 \\ 0 & 0 & 1 \end{bmatrix}$$

We get

$$A \times_n \mathbf{v} = \begin{bmatrix} v^2 \\ v^1 \\ v^3 \end{bmatrix}$$

$$A \times_n \mathbf{w} = \begin{bmatrix} w^2 \\ w^1 \\ w^3 \end{bmatrix}$$

And

$$(A \times_n \mathbf{v}, A \times_n \mathbf{w}) = (A \times_n \mathbf{v})^T \times_n (A \times_n \mathbf{w}) =$$

$$= v^1 \times_2 w^1 +_2 v^2 \times_2 w^2 +_2 v^3 \times_2 w^3 = (\mathbf{v}, \mathbf{w})$$

i.e.

$$((A \times_n \mathbf{v})^T, (A \times_n \mathbf{w})^T) = (\mathbf{v}^T, \mathbf{w}^T)$$

Generally speaking we have

$$(A \times_n \mathbf{v}, A \times_n \mathbf{w}) = (A \times_n \mathbf{v})^T \times_n (A \times_n \mathbf{w}) = (\mathbf{v}^T \times_n A^T) \times_n (A \times_n \mathbf{w}) =$$

$$= \mathbf{v}^T \times_n (A^T \times_n A) \times_n \mathbf{w} +_n \eta_1 = \mathbf{v}^T \times_n \mathbf{w} +_n \eta_1$$

where η_1 is a random variable depends on n and m. That means the set of orthogonal matrices does not preserve Euclidean metric in $E_3 W_n$, but preserves it with some probability less than 1.

Let's again takes $n = 2$, orthogonal matrix A:

$$A = \begin{bmatrix} 0.81 & 0.69 & 0 \\ -0.69 & 0.81 & 0 \\ 0 & 0 & 1 \end{bmatrix}$$

and calculate

$$A \times_2 A = \begin{bmatrix} 0.28 & 0.96 & 0 \\ -0.96 & 0.28 & 0 \\ 0 & 0 & 1 \end{bmatrix}$$

So, we get

$$A \times_2 A \notin \mathbf{O} \ (3)$$

Let's again takes $n = 2$, another orthogonal matrix A:

$$A = \begin{bmatrix} 0 & 1 & 0 \\ -1 & 0 & 0 \\ 0 & 0 & 1 \end{bmatrix}$$

and calculate

$$A \times_2 A = \begin{bmatrix} -1 & 0 & 0 \\ 0 & -1 & 0 \\ 0 & 0 & 1 \end{bmatrix}$$

We get

$$A \times_2 A \in \mathbf{O} \ (3)$$

Generally speaking, the set $\mathbf{O}$ (3) is not the Lie group in Mathematics with Observers, and group definition's conditions take a place here with some probability less than 1.

So, we proved

Theorem 12.2 *The set O (3) is not the Lie group in Mathematics with Observers, and group definition's conditions take a place here with some probability less than 1.*

Let's consider now four-dimensional Minkowski space M^4 with real coordinates x^μ for $\mu = 0, 1, 2, 3$, equipped with the Minkowski metric as a matrix

$$(\eta) = (\eta)_{\mu\nu} = (\eta)^{\mu\nu} = \begin{bmatrix} 1 & 0 & 0 & 0 \\ 0 & -1 & 0 & 0 \\ 0 & 0 & -1 & 0 \\ 0 & 0 & 0 & -1 \end{bmatrix}$$

In classic Mathematics a Lorentz transformation is a linear transformation

$$\Lambda : M^4 \longrightarrow M^4$$

which transforms coordinates $x'^\mu = \Lambda^\mu_\nu x^\nu$ for $\Lambda^\mu_\nu \in \mathbf{R}$, but leaves the length invariant

$$(\eta)_{\mu\nu} x'^\mu x'^\nu = (\eta)_{\mu\nu} x^\mu x^\nu$$

for all x.

This condition can be rewritten in matrix notation as

$$\Lambda^T \eta \Lambda = \eta$$

Now we are going to Mathematics with Observers point of view. We consider the set $\mathbf{L}$ of all 4×4 matrices $\Lambda = (\Lambda^\mu_\nu)$, where $\Lambda^\mu_\nu \in W_n$, satisfying the condition

$$\Lambda^T \times_n \eta \times_n \Lambda = \eta$$

and call these matrices the Lorentz matrices.

Let's note we have in this case

$$(\Lambda^T \times_n \eta) \times_n \Lambda = \Lambda^T \times_n (\eta \times_n \Lambda)$$

So, the condition

$$\Lambda^T \times_n \eta \times_n \Lambda = \eta$$

has unique sense.

Now let's go to the set $\mathbf{L}$ of all Lorentz matrices.

Let's consider $n = 2$ and matrix Λ:

$$\Lambda = \begin{bmatrix} 1 & 0 & 0 & 0 \\ 0 & 0.81 & 0.69 & 0 \\ 0 & -0.69 & 0.81 & 0 \\ 0 & 0 & 0 & 1 \end{bmatrix}$$

We have

$$\Lambda^T \times_n \eta \times_n \Lambda = \eta$$

So, Λ is a Lorentz matrix, i.e.

$$\Lambda \in \mathbf{L}$$

But if we consider matrix

$$\Gamma = \Lambda \times_2 \Lambda = \begin{bmatrix} 1 & 0 & 0 & 0 \\ 0 & 0.28 & 0.96 & 0 \\ 0 & -0.96 & 0.28 & 0 \\ 0 & 0 & 0 & 1 \end{bmatrix}$$

we get

$$\Gamma^T \times_2 \eta \times_2 \Gamma = \begin{bmatrix} 1 & 0 & 0 & 0 \\ 0 & -0.85 & 0 & 0 \\ 0 & 0 & -0.85 & 0 \\ 0 & 0 & 0 & -1 \end{bmatrix}$$

i.e. in this case

$$\Gamma^T \times_n \eta \times_n \Gamma \neq \eta$$

So, Γ is not a Lorentz matrix, i.e.

$$\Gamma \notin \mathbf{L}$$

Let's note if we take any $n \geq 3$ and substitute matrix Λ elements 0.81 to 0.8$\underbrace{0...0}_{n-2}$1, ± 0.69 to $\pm 0.6 \underbrace{0...0}_{n-2}$9, we get the same result.

That means the set $\mathbf{L}$ is not a group, i.e. classic Lorentz Lie group is not a group in Mathematics with Observers, and group operation conditions are satisfied in the set $\mathbf{L}$ with some probability less than 1.

Theorem 12.3 *The set $\mathbf{L}$ is not a group in Mathematics with Observers, and group operation conditions are satisfied in the set $\mathbf{L}$ with some probability less than 1.*

The Poincare group in classic mathematics consists of Lorentz transformations combined with translations, which act on the spacetime coordinates by

$$x^\mu \longrightarrow \Lambda^\mu_\nu x^\nu + b^\mu$$

where Λ is a Lorentz transformation and $\mathbf{b} = (b^0, b^1, b^2, b^3) \in \mathbf{R}^4$

is an arbitrary vector. I.e. the Poincare group is the set of 5×5 matrices P of the form

$$P = \begin{bmatrix} \Lambda & \mathbf{b} \\ \mathbf{0} & 1 \end{bmatrix}$$

where Λ is a Lorentz transformation, $\mathbf{b}$ is an arbitrary column-vector $\in \mathbf{R}^4$ and $\mathbf{0}$ is a zero row-vector $\in \mathbf{R}^4$, under matrix multiplication:

$$P_1 = \begin{bmatrix} \Lambda_1 & \mathbf{b}_1 \\ \mathbf{0} & 1 \end{bmatrix}$$

$$P_2 = \begin{bmatrix} \Lambda_2 & \mathbf{b}_2 \\ \mathbf{0} & 1 \end{bmatrix}$$

$$P_1 \times P_2 = \begin{bmatrix} \Lambda_1 \times \Lambda_2 & \Lambda_1 \times \mathbf{b}_2 + \mathbf{b}_1 \\ \mathbf{0} & 1 \end{bmatrix}$$

Now we are going to Mathematics with Observers point of view. We consider the set $\mathbf{P}$ of all 5×5 matrices $P = (P_\nu^\mu)$, where $P_\nu^\mu \in W_n$, of the form

$$P = \begin{bmatrix} \Lambda & \mathbf{b} \\ \mathbf{0} & 1 \end{bmatrix}$$

where Λ is a Lorentz transformation, $\mathbf{b}$ is an arbitrary column-vector $\in E_4 W_n$ and $\mathbf{0}$ is a zero row-vector $\in E_4 W_n$, under matrix multiplication:

$$P_1 = \begin{bmatrix} \Lambda_1 & \mathbf{b}_1 \\ \mathbf{0} & 1 \end{bmatrix}$$

$$P_2 = \begin{bmatrix} \Lambda_2 & \mathbf{b}_2 \\ \mathbf{0} & 1 \end{bmatrix}$$

$$P_1 \times P_2 = \begin{bmatrix} \Lambda_1 \times_n \Lambda_2 & \Lambda_1 \times_n \mathbf{b}_2 +_n \mathbf{b}_1 \\ \mathbf{0} & 1 \end{bmatrix}$$

Now let's go to the set $\mathbf{P}$ of all Poincare matrices.
Let's consider $n = 2$ and matrix P:

$$P = \begin{bmatrix} 1 & 0 & 0 & 0 & 0 \\ 0 & 0.81 & 0.69 & 0 & 0 \\ 0 & -0.69 & 0.81 & 0 & 0 \\ 0 & 0 & 0 & 1 & 0 \\ 0 & 0 & 0 & 0 & 1 \end{bmatrix}$$

Here

$$\Lambda = \begin{bmatrix} 1 & 0 & 0 & 0 \\ 0 & 0.81 & 0.69 & 0 \\ 0 & -0.69 & 0.81 & 0 \\ 0 & 0 & 0 & 1 \end{bmatrix}$$

is a Lorentz matrix, $\Lambda \in \mathbf{L}$, $\mathbf{b} = \mathbf{0}$.

But if we consider matrix

$$Q = P \times_2 P = \begin{bmatrix} 1 & 0 & 0 & 0 & 0 \\ 0 & 0.28 & 0.96 & 0 & 0 \\ 0 & -0.96 & 0.28 & 0 & 0 \\ 0 & 0 & 0 & 1 & 0 \\ 0 & 0 & 0 & 0 & 1 \end{bmatrix}$$

where

$$\Gamma = \Lambda \times_2 \Lambda = \begin{bmatrix} 1 & 0 & 0 & 0 \\ 0 & 0.28 & 0.96 & 0 \\ 0 & -0.96 & 0.28 & 0 \\ 0 & 0 & 0 & 1 \end{bmatrix}$$

And we get

$$\Gamma^T \times_2 \eta \times_2 \Gamma = \begin{bmatrix} 1 & 0 & 0 & 0 \\ 0 & -0.85 & 0 & 0 \\ 0 & 0 & -0.85 & 0 \\ 0 & 0 & 0 & -1 \end{bmatrix}$$

i.e.

$$\Gamma^T \times_n \eta \times_n \Gamma \neq \eta$$

So, Q is not a Poincare matrix, i.e.

$$Q \notin \mathbf{P}$$

Let's note if we take any $n \geq 3$ and substitute matrix P elements 0.81 to $0.8\underbrace{0...0}_{n-2}1$, ± 0.69 to $\pm 0.6\underbrace{0...0}_{n-2}9$, we get the same result. That means the set $\mathbf{P}$ is not a group, i.e. classic Poincare Lie group is not a group in Mathematics with Observers, and group operation conditions are satisfied in the set $\mathbf{P}$ with some probability less than 1.

So, we proved

Theorem 12.4 *The set $\mathbf{P}$ is not a group in Mathematics with Observers, and group operation conditions are satisfied in the set $\mathbf{P}$ with some probability less than 1.*

As a result we get

Theorem 12.5 *In Mathematics with Observers the probabilities of the conservation laws characterizing special relativistic fluid mechanics are invariant (in fact co-variant) under Poincare group of transformations are less than 1.*

13

Appendix: Review of Publications of the Mathematics with Observers

This Appendix contains the review of authors's publications of the Mathematics with Observers in period from 2004 until 2020. This set of publications includes the foundations of Mathematics with Observers, problems of classic pure mathematics and problems of contemporary physics from Mathematics with Observers point of view.

The foundations of Mathematics with Observers are introduced in works "Boris Khots and Dmitriy Khots, Mathematics of Relativity – Observer's Mathematics, Web book, 2020, 2019, 2004, www.mathrelativity.com" and "Boris Khots, Dmitriy Khots, An Introduction to the Mathematics of Relativity, Lecture Notes in Theoretical and Mathematical Physics, Ed. A.V. Aminova, Kazan State University, v. 7, pp 269-306, 2006". We introduced here Observers, arithmetic operations depend on observers, considered the main properties of this new arithmetic. Also we considered the situations in algebra, geometry, mathematical analysis which appear there in connection with this arithmetic.

The problems of classic pure mathematics were considered in several publications. In works "Boris Khots, Dmitriy Khots, Analogy of Fermat's last problem in Observer's Mathematics – Mathematics of Relativity, Talk at the International Congress of Mathematicians, Madrid 2006, Proceedings of ICM2006" and "Dmitriy Khots, Boris Khots, Fermat's, Mersenne's and Waring's problems in Observer's Mathematics, International Journal of Pure and Applied Mathematics, Bulgaria, v. 43, 3, pp 403-408 (2008)" the following theorems were proved:

Theorem 13.1 *Analogue of Fermat's Last Problem. For any integer n, $n \geq 2$, and for any integer m, $m \geq 3$, $m \in W_n$ there are positive a, b, $c \in W_n$, such that $a^m +_n b^m = c^m$.*

Theorem 13.2 *Analogue of Mersenne's numbers problem. There are integers n, $k \geq 2$, Mersenne's numbers M_k, with $\{k, M_k\} \in W_n$, and positive $a \in W_n$, such that $M_k = a^2$.*

Theorem 13.3 *Analogue of Fermat's numbers problem. There are integers n, $k \geq 2$, Fermat's numbers F_k, $\{k, F_k\} \in W_n$, and positive $a \in W_n$, such that $F_k = a^2$.*

DOI: 10.1201/9781003175902-13

Theorem 13.4 *Analogue of Waring's problem. For any integer k, $k \geq 2$, there is integer n, $n \geq 2$, $(k \in W_n)$ and some $x \in W_n$ such that any equality of the form $x = a_1^k +_n a_2^k +_n \ldots +_n a_l^k$ is not possible for any integer $l \in W_n$ and any positive numbers $a_1, a_2, \ldots, a_l \in W_n$.*

In work "Boris Khots, Dmitriy Khots, Observer's Mathematics – Mathematics of Relativity, Applied Mathematics and Computations, volume 187, issue 1, April 2007, pp 228-238, New York" applications of Mathematics with Observers to mathematical analysis were considered.

In works "Dmitriy Khots, Boris Khots, Tenth Hilbert Problem in Observer's Mathematics, Proceedings of the 16th International Conference on Finite or Infinite Dimensional Complex Analysis and Applications (16th IC-FIDCAA), pp 81-85, Dongguk University, Gyeongju, Korea, 2008" and "Boris Khots, Dmitriy Khots, Analogue of Hilbert's tenth problem in Observer's Mathematics, Talk at the International Congress of Mathematicians, Hyderabad 2010, Proceedings of ICM2010" the following theorem was proved:

Theorem 13.5 *For any positive integers $m, n, k \in W_n$, $n \in W_m$, $m > log_{10}(1 + (2 \cdot 10^{2n} - 1)^k)$, from the point of view of the W_m–observer, there is an algorithm that takes as input a multivariable polynomial $f(x_1, \ldots, x_k)$ of degree q in W_n and outputs YES or NO according to whether there exist $a_1, \ldots, a_k \in W_n$ such that $f(a_1, \ldots, a_k) = 0$.*

In works "Dmitriy Khots, Euclidean and Lobachevsky Geometries in Mathematics of Relativity, Recent Problems in Field Theory, Ed. A.V. Aminova, Kazan State University, v. 5, pp 243-246, 2006", "Dmitriy Khots, Boris Khots, Non-Euclidean Geometry in Observer's Mathematics, Acta Physica Debrecina, tomus XLII, pp 112-119, Debrecen, Hungary, August 2008" and "Dmitriy Khots, Boris Khots, Solitary Waves and Dispersive Equations from Observer's Mathematics point of view, Geometry in large , topology and applications, pp 86-95, Kharkov, Ukraine,2010" the following theorem was proved:

Theorem 13.6 *Fix the x-axis, l_0, and pick a point on the y-axis, say $(0, b)$. Then the line*

$$y = k \times_n x +_n b$$

is parallel to l_0 in Lobachevsky sense iff $|b| \geq 1$, and in case $|b| < 1$, we would only have parallel lines in Euclidean sense.

The work "Boris Khots, Dmitriy Khots, Observer's Mathematics applications to Number Theory, Geometry, Analysis, Classical and Quantum Mechanics, Scientific Notes of Kazan State University, Physics – Mathematics series, 2011, v. 153, number 3, pp 196-203" contains the overview of applications of Mathematics with Observers to classic mathematics problems.

The problems of Quantum Mechanics from Mathematics with Observers point of view were considered in several publications. These are works: "Boris Khots, Dmitriy Khots, Quantum Mechanics from Observer's Mathematics point of view, 132 pp, DOI: 10.12732/acadpublmon2015007, Academic Publications, 2015", "Boris Khots, Dmitriy Khots, Small deviations between classical and observer's mathematics point of view on quantum mechanics, 16 pp, Fields Institute, Toronto, Canada, March 2015", "Boris Khots, Dmitriy Khots, Classical and Quantum Mechanics aspects from Observer's Mathematics point of view, Talk at the International Congress of Mathematicians, Seoul 2014, Proceedings of ICM2014", "Dmitriy Khots, Boris Khots, What is Behind Small Deviations of Quantum Mechanics Theory from Experiments? Observer's Mathematics Point of View, American Institute of Physics, Melville, New York volume 1637, 491 (2014)", "Boris Khots, Dmitriy Khots, Observer's Mathematics applications to the Quantum Mechanics, The Royal Swedish Academy of Sciences, Physica Scripta, volume 2014, T163, December 2014". There are several other publications with Mathematics with Observers applications to Quantum Mechanics. In particular we considered Dirac equation for free electron. We proved that this equation becomes stochastic and contains random variables. We considered so called Nadezhda effect – the distance between two different points does not always exist, i.e. not every segment on a space has a length. This phenomenon occurs for all W_n. This effect gives us new possibilities for discovering physical processes and developing their mathematical models.

In 1922, Albert Einstein received the Nobel Prize – not for his relativity theory, but for his interpretation of the photoelectric effect as being due to particle-like photons striking the surfaces of metals and ejecting electrons. But ironically it has been cogently argued that Einstein's conclusions were not fully justified. The theory of Lamb and Scully treated atoms quantum - mechanically, but regarded light as being a purely classical electromagnetic wave with no particle properties. Their conclusion was that the photoelectric effect does not constitute proof of the existence of photons. In work "Dmitriy Khots, Boris Khots, Photoelectric Effect from Observer's Mathematics Point of View, American Institute of Physics, Melville, New York, volume 1637, 487 (2014)" we showed that with enough small intensities Einstein interpretation of the photoelectric effect as being due to particle-like photons striking the surfaces of metals and ejecting electrons is correct.

The problems of Special and General Relativity theory from Mathematics with Observers point of view were considered in several publications. These are works: "Boris Khots and Dmitriy Khots, Special and General Relativity theory and Gravitation from Observer's Mathematics point of view, 120 pp, ISBN 978-5-906818-47-8, KURS Publishing House, 2016", "Boris Khots, Dmitriy Khots, Special Relativity from Observer's Mathematics point of view, Proceedings of the SPIE Optics + Photonics 2015 Conference 570 – The Nature of Light: What are Photons? VI, vol. 9570, pp 95701E-1 – 9570E-12, DOI: 10.1117/12.2185509, San Diego, California, USA, 2015", "Boris Khots, Dmitriy Khots, Lagrangian in Classical Mechanics and in Special Relativity from Observer's Mathematics Point of View, DOI 10.1007/s10701-015-9895-4, Foundations of Physics ISSN: 0015-9018 (Print) 1572-9516 (Online), Springer,

March 2015, Heidelberg, Germany; New York, USA", "Boris Khots, Dmitriy Khots, Hamilton equations of general relativity in Observer's Mathematics, Proceedings of 4th Chaotic Modeling and Simulation (CHAOS 2011) International Conference, Agios Nikolaos, Crete Greece, pp 203-208, 2011". In particular, Lorentz transformation becomes stochastic, tensor's nature of General Relativity transfers to the special basis.

The problems of Electrodynamics and Thermodynamics theory from Mathematics with Observers point of view were considered in work "Boris Khots and Dmitriy Khots, Electrodynamics and Thermodynamics from Observer's Mathematics point of view, 144 pp, ISBN 978-5-906923-68-4, KURS Publishing House, 2017". Maxwell equations become stochastic and contain random variables and random vectors.

The problems of Standard Model of particle physics from Mathematics with Observers point of view were considered in work "Boris Khots and Dmitriy Khots, The spin degree of freedom from Observer's Mathematics point of view, Proceedings Volume 11470, Spintronics XIII; 114703W (2020), https://doi.org/10.1117/12.2567105. Event: SPIE Nanoscience + Engineering, 2020, Online Only". The following theorem was proved:

Theorem 13.7 *In Mathematics with Observers the probability of spin* $- j$ *transformation is a homomorphism (representation) of* $SU(2)$ *to the set of matrix transformations of a linear space of polynomial functions is less than 1.*

And following by this theorem the connections between elementary fermions and half-integer spin and also between elementary bosons and integer spin were considered in Mathematics with Observers.

14

Glossary

Equation of continuity in Mathematics with Observers: (see Chapter 7):

$$\partial\rho/\partial t +_n \rho \times_n div\mathbf{v} +_n (\mathbf{v}, \mathbf{grad}\rho) +_n \omega_4 = \omega_3$$

Euler equation of the fluid in Mathematics with Observers: (see Chapter 8):

$$\partial\mathbf{v}/\partial t +_n (\mathbf{v}, \nabla) \times_n \mathbf{v} +_n \omega_7 = -\mathbf{grad}h +_n \mathbf{g} +_n \eta_{21}$$

Energy and moment flux in Mathematics with Observers:

$$(\text{see Chapter 9}) : \partial/\partial t(\frac{1}{2} \times_n (\rho \times_n (\mathbf{v}, \mathbf{v})) +_n \rho \times_n \epsilon) =$$

$$= -(\frac{1}{2} \times_n (\mathbf{v}, \mathbf{v})) \times_n div(\rho \times_n \mathbf{v}) -_n (\mathbf{v}, \mathbf{grad}\ p) -_n (\rho \times_n \mathbf{v}, (\mathbf{v}, \nabla) \times_n \mathbf{v}) -_n$$

$$-_n div(\rho \times_n \mathbf{v}) \times_n \epsilon +_n \rho \times_n \partial\epsilon/\partial t +_n \omega_{16}$$

and

$$\partial/\partial t(\int_e^g \int_c^d \int_a^b ((\rho \times_n v_i \times_n \Delta x_1) \times_n \Delta x_2) \times_n \Delta x_3) =$$

$$= -(\int_e^g \int_c^d (\Pi_{i1}(b, x_2, x_3) \times_n \Delta x_2) \times_n \Delta x_3 -_n$$

$$-_n \int_e^g \int_c^d (\Pi_{i1}(a, x_2, x_3) \times_n \Delta x_2) \times_n \Delta x_3 +_n$$

$$+_n \int_e^g \int_a^b (\Pi_{i2}(x_1, d, x_3) \times_n \Delta x_1) \times_n \Delta x_3 -_n$$

$$-_n \int_e^g \int_a^b (\Pi_{i2}(x_1, c, x_3) \times_n \Delta x_1) \times_n \Delta x_3 +_n$$

$$+_n \int_a^b \int_c^d (\Pi_{i3}(x_1, x_2, g) \times_n \Delta x_2) \times_n \Delta x_1 -_n$$

$$-_n \int_a^b \int_c^d (\Pi_{i3}(x_1, y_2, e) \times_n \Delta x_2) \times_n \Delta x_1) +_n \omega_{23}^i$$

DOI: 10.1201/9781003175902-14

Navier-Stokes equations in Mathematics with Observers: (see Chapter 11): For incompressible fluid

$$\begin{cases} \rho \times_n (\partial v_i/\partial t +_n \sum_{j=1}^3 {}^n v_j \times_n \partial v_i/\partial x_j) = -\partial p/\partial x_i +_n f_i(\mathbf{x},t) +_n A_i \\ A_i = \alpha \times_n (\sum_{j=1}^3 {}^n \partial^2 v_i/\partial x_j^2) +_n \omega_{29}^i +_n \omega_{26}^i \\ div\mathbf{v} = \sum_{j=1}^3 {}^n \partial v_j/\partial x_j = \omega_{28} \\ i = 1,2,3 \\ \alpha > 0 \end{cases}$$

Solutions of Navier-Stokes equations (see Chapter 11): For any not random divergence-free initial conditions, for any not random externally applied force Navier-Stokes equations for incompressible fluid in Mathematics with Observers do not have any not random solutions, i.e. we get solutions breakdown of Navier-Stokes equations in Mathematics with Observers for 3-dimensional case.

Bibliography

[1] A. Bertozi and A. Majda. *Vorticity and Incompressible Fluid.* Cambridge University Press, 2002.

[2] Jean Alexander Dieudonne. *Foundation of Modern Analysis.* Academic Press, 1969.

[3] D. Hilbert. *The Foundation of Geometry (Translation from German).* The Open Court Publishing Company, 1950.

[4] B. Khots and D. Khots. An introduction to mathematics of relativity. In A.V. Aminova, editor, *Lecture Notes in Theoretical and Mathematical Physics*, volume 7, pages 269–306, 2006.

[5] B. Khots and D. Khots. *Mathematics of Relativity (Observer's Mathematics).* Web Book, www.mathrelativity.com, 2020.

[6] L.D. Landau and E.M. Lifshitz. *The Classical Theory of Fields (Translation from Russian), Course of Theoretical Physics, volume 2,.* 1972.

[7] L.D. Landau and E.M. Lifshitz. *Fluid Mechanics (Translation from Russian), Course of Theoretical Physics, volume 6,.* 1987.

[8] P.K. Rashevsky. *Riemannian Geometry and Tenzor Analysis (in Russian).* 1955.

[9] P.K. Rashevsky. On the dogma of positive integers. In *Russian Math Survey*, volume 28, pages 246–249, 1973.

[10] Daniel V. Schroeder. *Thermal Physics.* Addison Wesley Longman, 2000.

[11] Steven Weinberg. *The Quantum theory of fields.* Cambridge University Press, 1995.

Index

Printed in the United States
by Baker & Taylor Publisher Services